PREFACE

During her short-lived respite from foreign domination, dating from the end of British colonial rule in 1960 to the devastating Turkish invasion in 1974, the Republic of Cyprus enjoyed an unprecedented interlude of political self-determination and economic prosperity. This atmosphere of economic growth pervaded the internal agricultural marketing system of Cyprus in 1970. The availability of a complete set of origin-destination produce flow data for the calendar year 1970 was propitious. It provided a rare opportunity to examine the transactional organization and spatial structure of an urban wholesale marketing system functioning within a rapidly developing Middle East economy.

By way of overview, Chapter I introduces the context of the study by discussing previous literature on internal marketing system development and the insights provided by agricultural location theory. With this theoretical background in mind, the salient demographic and economic aspects of Cyprus are outlined, and the strategy for answering the research questions is presented. Chapter II deals with the transactional organization of wholesalers in the marketplace. The stage is set by describing the situation and site of the Nicosia Municipal Wholesale Market. Next, the role of urban wholesalers in the marketing chain is discussed. This preliminary section is followed by a prinicipal components analysis designed to identify shared patterns of flow linkage between wholesaler destinations in the central market and supply origins in the hinterland. Chapter III focuses on the spatial structure of supply in the hinterland. The general pattern of produce flows to the central market is depicted and the dominant theme discussed. Next, the principal determinants of regional specialization in the hinterland are considered by singling out the two leading value of flow commodities for detailed analysis. Environmental and market distance variables are then entered into a multiple regression model to assess the relative importance of the Ricardo and Thünen approaches to horticultural location. Finally, in Chapter IV, the major findings of the study are summarized and related to the body of literature discussed in Chapter I.

Collection of the data base and marketplace surveys was undertaken the summer of 1974. Our field study was abruptly terminated by the July 20, 1974, Turkish invasion of Cyprus.[1] During the emergency evacuation of our home in Aglandja, six steamer trunks filled with market research data were left behind in the path of the advancing Turkish forces. A special debt of gratitude is due John, Ellie, and Michael Lysandrou for subsequently retrieving the data from the embattled area around our home. They then completed the tedious job of codifying the market records in their own home, and sent them on to us in the United States. We were thereby able to resume data processing at the University of Chicago in January of 1975.

Throughout the field study phase of our research, the Greek Cypriot people extended their unparalleled hospitality and cooperation. Tassos Anastassiades, Senior Economist in the Ministry of Finance, provided crucial liason with key administrative officers in the marketing system. His guidance and advice were essential to the completion of the study. P. Mallis, Chief Statistician in the Ministry of Finance, suggested sampling procedures at the inception of the study, gave us access to commodity trade summaries compiled by the Statistics and Research Department, and provided a staff of six government keypunch operators to process the codified market records. S. Ferekides, Inspector of Wholesale Markets, explained marketplace operation, allowed us to survey the marketplace sellers and buyers, and most importantly, permitted us to borrow a six-foot high stack of daily market receipts covering the year 1970. A. Christofi, Director of the Department of Lands and Surveys, advised on cartographic analysis. C. V. Economides, N. Vakis, and Chr. Olymbios, all on the staff of the Agricultural Research Institute, provided valuable insight into the locational determinants of produce supply areas. Leondios Hjiparaskevas, Marketing Officer in the Ministry of Commerce and Industry, shared his in-depth knowledge of the internal marketing system with us. Both Leondios Hjiparaskevas and Costas Apostolides made constructive comments on the manuscript.

During the latter stages of my graduate work at the University

---

[1]Appendix A illustrates the extent of the Turkish occupied zone as it exists today, and indicates the name and location of twenty-two sample produce supply villages and towns which are now truncated from the central wholesale market by the Attila Line.

[redacted]

FRESH FOOD FOR NICOSIA

by

Edward A. Kanne
Sears, Roebuck and Co.

THE UNIVERSITY OF CHICAGO
DEPARTMENT OF GEOGRAPHY
RESEARCH PAPER NO. 190

1979

Library of Congress Cataloging in Publication Data

Kanne, Edward A. 1943–

Fresh food for Nicosia

(Research Paper—University of Chicago, Department of Geography; no. 190)

Bibliography: p. 97

1. Produce trade—Cyprus—Nicosia. 2. Wholesale trade—Cyprus—Nicosia. I. Title. II. Series: Chicago. University. Dept. of Geography. Research Paper; No. 190.

HD 9016. C93N524 381′ .41′ 095645

ISBN 0-89065-097-7

79-10709

Research Papers are available from:

The University of Chicago
Department of Geography
5828 S. University Avenue
Chicago, Illinois 60637
Price: $6.00 list; $5.00 series subscription

of Chicago, I benefited from the joint advisement of Professors Marvin Mikesell and Brian Berry. I am particularly grateful to them for their timely encouragement. Brian Berry made crucial editorial and analytic comments on an earlier draft of the manuscript.

I owe a special debt of gratitude to my daughters Zacho and Allena for their understanding, and to my wife Koula for her staunch support and assistance in the research and writing of this paper.

E. A. Kanne
Corporate Store Location Research
Sears Tower
Chicago, Illinois
November, 1978

TABLE OF CONTENTS

LIST OF TABLES

LIST OF ILLUSTRATIONS

CHAPTER I

INTRODUCTION

Context of the Study

A current consensus of opinion among leading food and agricultural specialists is that the most effective means of controlling the synergic problems of low food supplies, high population growth rates, and abject poverty in agrarian societies is to increase indigenous food production and profitability among small landholding farmers:

> An all-out effort to increase food production in the poor, food-deficit countries may be the best means of raising incomes and accumulating the capital for economic development, and thus for moving the poor countries through the demographic transition to moderate rates of population growth (Wortman 1976, p. 31).

Three approaches to improving the productivity and income of farmers with tiny family-operated holdings have been given considerable attention recently. Wortman (1976) suggests the introduction of new high-yielding crop production technology tailored specifically to fit unique local ecosystems. Schultz (1964) advocates capital investment in agricultural training programs to enhance smallholders' skills. And, Dumont (1968;1973) arguing that genuine development will come through hand plows and donkey carts, points out that "indigenous agricultural systems can contribute to relief of poverty and provide the basis for authentic development in the underdeveloped world, but only if governments are prepared never to turn their backs on the interests of small farmers" (de Souza 1974, p. 6).

In most of the developing world, small-scale farming is carried on more to provide food for the family, or for the village community or tribe, than it is for sale to an outside market. Mosher (1969, p. 6) observes that if this situation continues, very little agricultural development can occur, and the levels of living of the people on these farms cannot rise:

> For agricultural development to take place, it is essential, among several other necessities, that farms become less and less subsistence and

> more and more commercial, producing increasingly for the market.

With specific reference to the Middle East region, McMillan (1969) feels that while efforts to increase agricultural production must be continued, further attention is needed to the development of modern, efficient systems for the marketing of this agricultural production. He stresses that developments in this area are necessary not only because marketing provides the essential link between the producer and the consumer, but it also has a major impact on the producer:

> Without accessible markets which provide him with a reasonable return for his efforts, there is little incentive for him to expand production beyond the needs of his family. From the standpoint of a country's overall economic development, efforts to expand and improve both the industrial and agricultural sectors, as well as efforts to improve the living conditions and nutritional standards of the population as a whole, require special attention to the development of efficient marketing for food and agricultural products (McMillan 1969, p. 66).

To recapitulate our position, if increased production is to bring higher income to small farmers, a critical shift in orientation must occur from local subsistence to commercial cash crop cultivation in response to greater domestic demand focused largely in the growing urban market centers. And, expansion of the internal marketing system provides an essential outlet for an increased intensity of cash crop farming (Belshaw 1965). Thus, the transition from subsistence to market-oriented production for domestic consumption requires a concomitant restructuring of marketing channels, transport networks, and marketplace facilities such that the internal marketing system is capable of moving increased flows of staple food crops efficiently from small-scale producers to urban centers of demand.

In view of its critical role in stimulating the productivity and income of small farmers in agrarian societies, what is the state of our knowledge regarding the structure and function of internal food marketing?

Internal Marketing of Staple Food Crops

The study of food commodity flows through internal marketing systems of the Third World has been neglected. De Souza and Porter (1974) emphasize that Western scientists, as well as government officials in many underdeveloped countries, know much more about how export crops move within colonial transportation networks and through port cities than they know about the marketing of food products for consumption in underdeveloped countries. The lacuna is expressed by Bohannan and Curtin (1971, p. 162):

> The amount of internal trade in various African countries that goes through the market place is tremendous, but nobody has any idea how much volume or what value of goods may actually be distributed.

Similarly, Jones (1972, p. 3) was appalled by how little quantitative information is available about the marketing of staple foods in tropical Africa:

> Almost nothing is known about the magnitude of the trade, very little about the distances foodstuffs travel, or even about the locations of the principal sources of some major foodstuffs.

The bulk of the literature dealing with the internal marketing of staple foods in developing countries has been primarily concerned with retail trade in rural periodic markets functioning within peasant economies (Bromley 1971; R. H. T. Smith 1972a, b; C. A. Smith 1974). Many of these studies have drawn conceptual impetus from central place theory applied to peasant market organization in Skinner's "Marketing and Social Structure in Rural China" (1964-5), and Berry's *Geography of Market Centers and Retail Distribution* (1967).

Hodder and Ukwu (1969) formulated a general conceptual framework of stages in the development of markets in Africa and Europe. On a continuum ranging from the most elementary to the most complicated economies, they trace the development of markets from the simplest rural periodic markets through to the daily urban markets consisting of retail shops and large specialized wholesale markets in the larger towns.

As a result of their examination of the relationship between evolving internal production and marketing systems for food staples in Northeast Brazil, Forman and Riegelhaupt (1970) propose a widely recognized (C. A. Smith 1974) standard sequence of marketing system development, ranging from spatially dispersed labor intensive local markets in predominance, to centralized capital intensive urban markets. The key mechanism of transition is the increasing degree of urban wholesaler dominance of trade in the marketing chain through highly capitalized economies of scale. In the advanced stages, a high degree of wholesaler concentration exists such that a disproportionately large share of trade value is accounted for by the largest firms (Metcalf 1970).

The changing food marketing system of the United States has evolved through three stages, the first two of which encompass the basic concepts behind Hodder and Ukwu's scheme for Africa and Europe, and Forman and Rieglehaupt's model for Brazil. The evolution of the system is a function of developments in transportation and communication as they affect the location and scale of the markets. The three stages, quoting from Shepherd and Futrell (1970), are:

1. Local markets (team and wagon)--small in size, large in number, located in small country towns. Direct producer to consumer retail

2. Central markets (railroad)--large in size, few in number, located at large cities. Bulk of the produce flows consigned to commission wholesalers on behalf of farmers and local dealers

3. Decentralized markets (truck, paved highways, and radio)--moderate size, growing number, chain store warehouse facilities located on the periphery of cities. Buyers and sellers short-cut the central market, transacting directly at "interior markets" (1970, pp. 78-80, 473).

As evinced in the studies of Hodder and Ukwu (1969), Forman and Rieglehaupt (1970), Bromley (1971), Jones (1972;1974), Norton and Symanski (1975), and V. Smith (1975), the spatial structure of the produce marketing systems of many developing economies has evolved out of the local market stage toward the central market stage in which urban wholesalers play a dominant role in terms of the volume and value of food bulked from farmers and local dealers and wholesaled to urban retailers:

> Wholesale assembly and distribution particularly is a key activity in the whole marketing system, since the wholesale market represents the focal point in the flow of goods from producer to consumer (U.S. Department of Agriculture 1954, p. 282).

Central urban wholesale markets and their food supply hinterlands have not received the detailed analysis which their growing importance in the internal marketing systems of dynamic economies warrants. Morevover, due largely to the lack of hard data,[1] few studies which have focused explicitly on urban wholesale food markets and hinterlands have used origin-destination commodity flow

---

[1]For discussions of the shortage of good commodity flow data in spatial interaction research see Isard (1960, p. 143), and South (1972, pp. 10-12).

analysis to reveal the spatial structure of the system. Of the four exemplary works using the spatial structure approach to commodity flow analysis (R. H. T. Smith 1970), three have analyzed volume of rail shipments data (Ullman 1957; Berry 1966; Reed 1967), and one has dealt with truck movements (Helvig 1964).[1]

Olson (1968), Gregor (1957), and Jumper (1969) have examined the volume of truck unloads of fresh produce at urban centers in the United States. However, no study, to date, has used actual value of flow data to elucidate the spatial interactions between a central wholesale market and its fresh food hinterland in a developing Middle East economy.

Some basic insights into how this gap in the literature might be filled are offered in classical agricultural location theory.

Agricultural Location Theory and Central Urban Markets

Historically, two variables have largely determined the spatial organization of agricultural supply hinterlands for central urban markets: (1) land productivity, and (2) market accessibility. Each of these variables has been given thorough consideration in an established body of theory developed in the early nineteenth century by Ricardo (1817) and Thünen (1826).

In essence, classical agricultural location theory posits that spatial variations in agricultural land use arise from differential land rents (i.e., return on investment per unit land) which, in turn, are due to soil productivity[2] differences (Ricardo 1817) and/or locational differences in distance to market (Thünen 1826) under conditions of a competitive market economy. Both Ricardo's and Thünen's explanations are held to be "central to the theory of agricultural location, in which the basic idea is that each parcel of land is used for the activity that can pay the highest rent"

---

[1]Following Berry's (1966) prototype "factor" analysis of Indian commodity flows, a number of geographers have applied the "principal components" model used in our analysis to a variety of flow matrices. See, for example, Goddard (1970), Britton (1971), Greer-Wooten (1972), South (1972), Black (1973), Schwind (1975), and Wiseman (1976).

[2]In current usage, soil <u>fertility</u> refers to the inherent capacity of a soil to supply nutrients to plants in adequate amounts and in suitable proportions. <u>Productivity</u> is related to the ability of a soil to yield crops. Productivity is the broader term since fertility is only one of a number of factors that determine the magnitude of crop yields (Buckman and Brady 1968, p. 7).

(Berry, Conkling, and Ray 1976).

In his second edition of The Isolated State, Thünen asks the reader to imagine a very large town at the center of a plain, the soil of which is capable of cultivation and of the same fertility throughout. The question he asks is "How will the farming system of different districts be affected by their distance from the Town?" He concludes:

> It is on the whole obvious that near the Town will be grown those products which are heavy or bulky in relation to their value and hence so expensive to transport that the remoter districts are unable to supply them. Here too we shall find the highly perishable products, which must be used very quickly. With increasing distance from the Town, the land will progressively be given up to products cheap to transport in relation to their value.
>
> For this reason alone, fairly sharply differentiated concentric rings or belts will form around the Town, each with its own particular staple product.
>
> From ring to ring the staple product, and with it the entire farming system, will change; and in the various rings we shall find completely different farming systems (Hall 1966, p. 8).

Note particularly that within Thünen's first ring, market-gardening occupies the land immediately around the central urban market:

> Delicate horticultural products such as cauliflower, strawberries, lettuce, etc., would not survive the long journeys by wagon. They can, moreover, be sold only in small quantities, while still quite fresh. All these products will be grown near the Town (Hall 1966, p. 9).

Thus in Thünen's original model, the key variable is distance to market. Transport costs reflect the expense of overcoming the friction of distance as well as the bulk, weight, and perishability of the commodities sent to market. And, as transport costs increase with distance, they impart a concentric "distance-decay" zonation of land rents in which intensity of cultivation declines with increasing distance from the central urban market.

Despite the simplifying assumption in his original paradigm, Thünen was fully cognizant of spatial variations in soil productivity in the real world:

> Nowhere in reality do we find soil of the same physical quality and of the same level of fertility throughout an entire country (Hall 1966, p. 171).

Given the assumption of uniform soil productivity and the basic premise that transport costs are the primary determinant of land use patterns, the logic of Thünen's distance-decay concentric zonation is irrefutable. Indeed, the confirmations of his theory are legion, particularly as it applies to areas in the early stages of economic development. However, there is general agreement that the pattern disintegrates as the economy of an area advances (Gregor 1970, p. 69) and some have gone so far as to argue for an inverted zonation (Sinclair 1967).[1]

Yet, much of this apparent inconsistency may be alleviated by going back to the original model, as Thünen might have done, and adjusting the assumptions and premises to fit the contemporary scene by allowing soil productivity to become spatially variant, and substituting production costs for transport costs as the principal control of agricultural location. The result is a pattern of regional specialization in areas possessing greatest natural advantage, and whether those special environmental conditions be found near the central urban market or far away becomes less relevant, even on a national scale.[2] In this modified version, if soil productivity is allowed to increase with distance from market, and transport costs are significantly reduced below production costs, the a "distance-growth" pattern of land use intensity may be derived.

Theodore Brinkman (1935) elaborates on the classic Thünen model by introducing the modernizing impact of increasing consumer demand for agricultural products (a function of urban population growth and rising per capita income), and concomitant improvements in transport technology. The spatial consequences of these facets of economic development are twofold: (1) an extension of the margin of intensive cultivation at greater distances from the central urban market, and (2) a related increase in the degree of regional

---

[1]Studies of the fresh produce supply areas of Tokyo (Eyre 1959) and Los Angeles (Gregor 1957) indicate that the location of a large urban market in the midst of an ideal physical environment helps to make the peri-urban supply area a leader in the production of fresh produce. However, Sinclair (1967) argues that when land rents for urban uses exceed those of agriculture, then urban encroachments becomes an important force in relocating intensively cultivated crops in environmentally favorable areas at greater distances from the central urban market. The former situation supports the distance-decay zonation of land rents in Thünen's paradigm, while the latter may generate an inverted or reverse zonation.

[2]This is essentially the conclusion that O. E. Baker (1921) arrived at in his seminal work on the role of physical conditions in determining agricultural land use patterns in the United States. See Gregor (1970) for a review of Baker's subsequent elaborations on this thesis.

specialization in the supply hinterland. Improvements in transport technology, particularly the extension of overland transport systems into rural areas, coupled with the use of cheaper and faster modes of transport, proportionately decrease market attraction for agricultural production, bringing a corresponding drop in the slopes of the rent-gradients.

Chisholm (1962, p. 167) elucidates the relationship between reduced transport costs and regional specialization of agricultural production:

> With a decline in transport costs, the advantage conferred by proximity to markets and to inputs diminishes in absolute importance. Consequently, other location factors must assume a larger relative significance, principally conditions of climate, soil and topography. Collectively, these have a big effect upon the earliness of the crop season and thus on the price which can be obtained for the produce and on the costs of production. The consequence is that instead of a particular crop being grown near the consuming centres and necessarily under very diverse physical conditions, it is increasingly possible for it to be grown in a limited number of places which physically are well suited to its requirement. The advantages of lower production costs and/or earlier season are not fully absorbed by transport costs, as previously they were.

In response to these contemporary locational forces, horticultural supply areas in particular have shifted significantly in both location and scale from urban-oriented market gardening of Thünen's time to modern resource-oriented truck farming.[1]

Supply Areas for Fresh Produce Sold in Urban Markets

Empirical literature on horticultural land use in North America and Western Europe corroborates the modern shift from market gardening to specialized truck farming in areas having particularly superior physical conditions (Conkling and Yeates 1976, p. 55).

In the United States, Jumper finds a rapidly declining role for the small market garden and an increasing significance for large-scale truck farms due largely to four factors:

1. The movement of suburbs onto urban fringe areas formerly devoted

---

[1]As in Thunen's original model, market gardens are found in the immediate environs of large cities and consist of small intensively cultivated plots producing a wide variety of vegetables for nearby urban markets. Truck farms, on the other hand, locate some distance from their markets, contain relatively large acreages, and specialize in a limited variety of vegetables (Jumper 1969, p. 311).

to market gardening--a trend that has been especially rapid since World War II and that has marked the permanent retirement from farming of many of the displaced farmers

2. Inability of small-scale methods of production to compete with larger more efficient truck farms irrespective of the market proximity of the larger units

3. Transportation improvements and the related decline in immediate proximity as a factor in lowering total production and marketing costs or in preventing deterioration of perishables and

4. Loss of the small farmer's local outlets for vegetables along with changes in methods of retail selling from produce stands and small independent grocers to chainstore supermarkets. Supermarket buyers seek sources able to supply dozens of carloads of vegetables rather than the dozens of bushels that normally mark the operation size of the vegetable farmer in the urban fringe (1969, p. 311).

Jumper concludes that an essential result of the decline in small-scale and fringe-area vegetable production is that the source of vegetables for urban consumers is moving further away.

A general rule of thumb seems to be that "large scale production of fresh vegetables occurs as close to major markets as environmental conditions and economic factors such as cost and availability of land and labor will permit" (Jumper 1969, p. 314).

The dynamic tension between Ricardo's and Thünen's explanations of agricultural land use is explicit in the cyclical advance and retreat of vegetable growing areas which supply the major northeast urban markets of the United States and Canada. Land productivity and market distance are engaged in a seasonal tug-of-war. Most of the high value early winter vegetables are grown in the frost-free southern extremities of the United States. As the frost-free season migrates northward in the spring, vegetable plantings follow, clustering in proximity to the northeast megalopolis in mid-summer, and returning to a southern exposure by autumn (Jumper 1969, p. 314). This seasonal migration of peak productivity occurs on both global (Whittlesey 1936) and local (Gregor 1957) supply scales.

California ranks highest in annual value of fresh vegetables sent to the major urban markets of the northeast, with Florida second, followed by Texas, and Arizona. Conkling and Yeates (1976) comment on the locational determinants of California's preeminence:

> Distance and land quality may exert opposing

> locational pressures in a given instance... as in the case of vegetable production in California. There the advantages of climatic and soil conditions outweigh the cost of shipping produce to markets in the northeastern United States.

Specific complementarity (Ullman 1957; R. H. T. Smith 1964) and interregional dependency (Jumper 1969) underlie the structure of fresh vegetable flows from the high value supply areas of the south and southwest to distant northeast urban markets where demand is strongest. On an annual value of flow basis and at a national scale, the dominant fresh vegetable supply areas in the United States are located:

1. along the <u>southern periphery of the land mass</u>, where the vegetables receive greater than 450 Langleys (gram calories per square centimeter) of mean annual solar radiation, and more than 2800 mean annual hours of sunshine
2. in <u>close proximity to the sea</u>, the meliorating influence of which provides a mean annual freeze-free period of greater than 300 days, particularly important for the production of high value early season crops grown under cover during March and April, and
3. on <u>irrigated lowlands</u>, rendering cultivation possible during the entire frost-free period. More than 80 percent of the productive land area of California and Florida is in gently sloping flat plains.

Thus, in general, comparative advantage in land productivity (especially climatic and hydrologic resources) plays a dominant role, and market distance (transport costs) plays a significant but subordinate role in determining the location of fresh vegetable supply areas in the United States.

To what extent do comparable land productivity variables explain the location of staple produce supply areas in Cyprus? Are basic underlying horticultural location forces at work in both the United States and Cyprus--despite the radically contrasting economic and areal scales?

In retrospect, the foregoing discussion of market system development and horticultural location theory lays the conceptual background against which to set our subsequent evaluation of the produce marketing milieu of Cyprus.

<u>The Cyprus Setting</u>

Cyprus is the third largest island in the Mediterranean after Sicily and Sardinia and covers an area of 3,572 square miles, about twice the size of Delaware. The shape of the island resembles a pegged-out lion's skin with its greatest length running 140 miles from the head in the southwest to the tip of the tail in the northeast; its greatest breadth from north to south is 60 miles. It is situated in the northeast corner of the Mediterranean Sea, 44 miles south of the Karamanian coast of Asia Minor, 60 miles west of the coast of Syria, and about 230 miles north of Egypt (Figure 1).

In 1970, Cyprus had an estimated population of 633,000, roughly that of Boston, Massachusetts. Three-fifths of the population was distributed among the 627 small nucleated rural villages, while the remaining two-fifths resided in the principal towns of Nicosia (the capital with 115,000 inhabitants), Limassol (51,500), Famagusta (42,500), Larnaca (21,400), Paphos (11,800), and Kyrenia (4,900). There were six administrative districts named after each of these towns.

The aggregate town population grew at an increasing rate, whereas village population experienced a decreasing rate of growth as a result of rural-urban migration (Cyprus 1970). During the 1946-1960 period, the urban population of Cyprus increased by 78 per cent with a corresponding increase of only 10 per cent in the rural sector, and the number of residents in the Nicosia suburbs alone jumped by 164 per cent (M. W. Hald 1968). This trend is reflected in a gradual reduction in the share of the labor force employed in agriculture from 40 per cent in 1946 to 35 per cent in 1971, together with compensatory increases in urban-oriented occupations. However, agriculture remains the backbone of the Cyprus economy, accounting for 20 per cent of the Gross Domestic Product in 1971 (Cyprus 1972c).

Increasing urbanization was accompanied by marked economic growth and rising levels of living during the period 1961-1971. Gross Domestic Product nearly doubled from £115 million in 1961 to £227 million in 1971 (i.e., at an annual growth rate of 7 per cent).[1] Improvements in the overall standard of living are indicated by an annual rate of increase of 6 per cent in both per capita Gross National Product and per capita private consumption expenditure. Further indications of a general rise in economic prosperity include significant improvements in health conditions, educational standards, housing facilities, and mass communications (Patsalides 1973).

---

[1]The Cyprus pound is equivalent to the British pound sterling, and is divided into 1,000 mils. In 1970, one Cyprus pound was equivalent to 2.6 United States dollars.

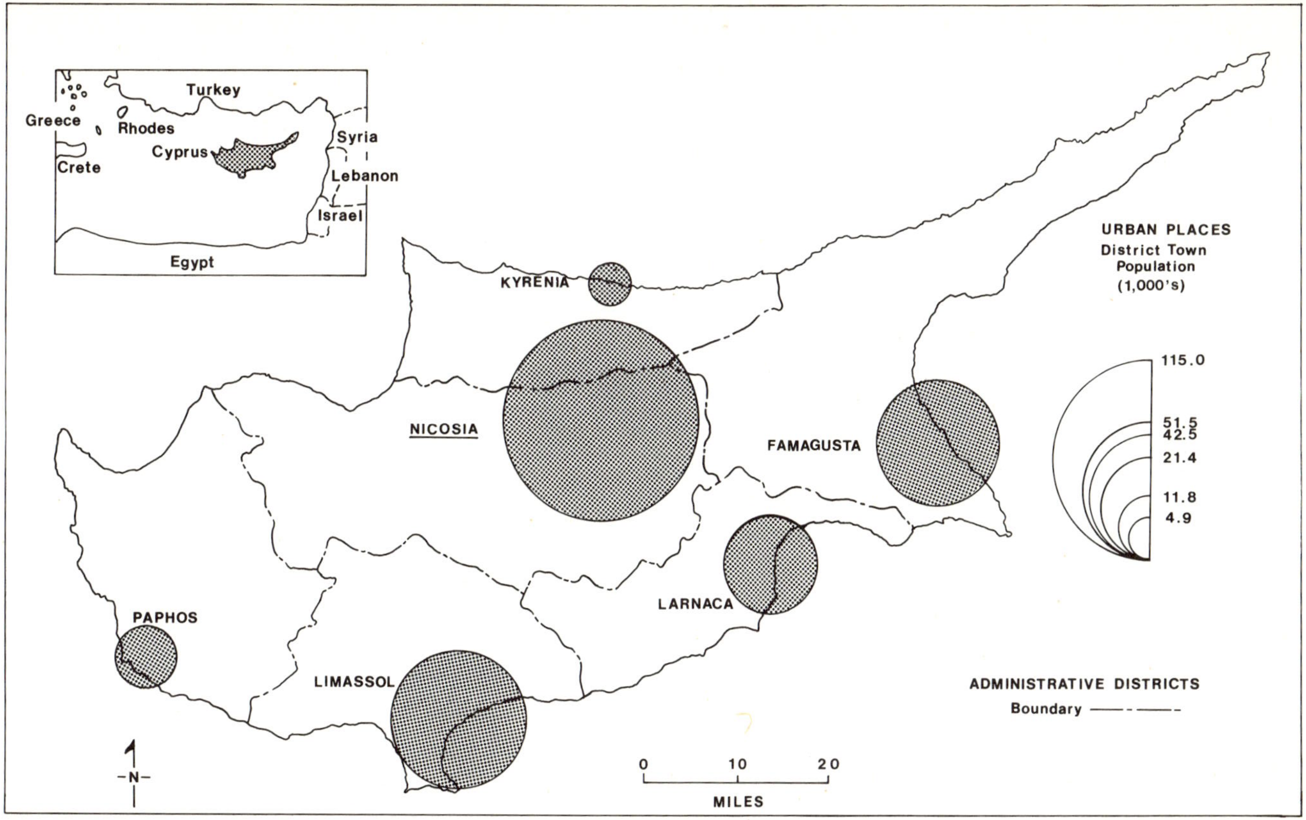

Fig. 1. Cyprus: urban places and administrative districts. Each town is the administrative center of the district which carries its name. The United Nations defines the population residing in the six district towns as urban. Insert shows the Island's position in the eastern Mediterranean.

The benefits of rising levels of living were not confined to the urban population, for Gross Agricultural Income shows an extraordinary rise of 129 per cent over the 1958-1971 period. Additional indices of agricultural development include increases in: amount of land under mechanical irrigation (from 53,131 donums[1] in 1946 to 213,505 donums in 1967); use of tractors on farms (from 120 in 1946 to 8,056 in 1971); and use of chemical fertilizers which has more than trebled since 1950 (Cyprus 1972a, 1969b, 1946b; Panagides 1967). Due largely to planned government expenditures in agricultural development programs, remarkable progress has been made in the modernization of traditional near-subsistence production and consequent expansion of the commercial farm sector.

Despite occasional setbacks due to the unpredictable scourge of drought (e.g., winter 1973), and the obdurate impediment of excessive land fragmentation (Karouzis 1970), Cyprus was becoming more and more self-sufficient in food production. Over the period 1967-1971, the island has shown progressive increments in a positive trade balance in the import/export ratio of food and live animals. Moreover, in 1970, 84 per cent of the total composition of private consumption expenditure on foodstuffs went to internally produced foods while only 16 per cent went toward the purchase of imported foodstuffs (Cyprus 1972c). Cypriot production of horticultural products has continued to satisfy the bulk of domestic consumer demands (Baldini 1968). With an average annual increase in food production of nearly 6 per cent, and a population growth rate of less than 1 per cent per annum, Cyprus ranked third (after Venezuela and Lebanon) among thirty developing nations whose rise in food production exceeded their growth in population and domestic demand for food for the period from 1953 through 1971 (Mayer 1976).

These economic developments, characteristic of the modernization of traditional Cypriot society, have wrought far-reaching changes in the internal produce market system. In short, as Cypriots migrated from rural agricultural villages to larger district towns, the remaining portion of the labor force employed in farming continued a gradual shift away from diversified subsistence production and local exchange, toward specialized commercial production and greater participation in national market exchange. Peasant subsistence consumption also declined because of the decreasing willingness of rural families, as their income rose, to produce and

---

[1]One Cyprus donum = 0.33 acre = 0.133 hectare.

process their own food. The combined effect of increasing urbanization, growing specialization in agriculture, and higher standards of living in villages and towns throughout the country has been to increase consumer demand for staple produce commodities (Table 1).

Extensive improvements in the transport network during the 1960-70 decade kept pace with growing consumer demand and the heavier volume of produce flows through marketing channels. The total length of bituminous surface public roads increased dramatically by 173 per cent, from 819 miles in 1960 to 2,234 miles in 1970. The main paved roads are well maintained by the Ministry of Communications and Works, are usable under all weather conditions, and reach all productive parts of the island. The secondary system of gravel roads totals 2,811 miles, most of which can be used by motor vehicles (Cyprus 1972c).

To sum up, there is in the Cyprus of 1970 an economic compage of a high level of consumer demand for staple produce commodities, combined with improved methods of cultivation in an increasingly specialized truck-farming hinterland which has excellent transport access to the central wholesale market at Nicosia.

Research Design

Having set the general context of the study, we now turn to the application of theories of urban wholesaling and hinterland specialization to the empirical produce marketing milieu of Cyprus. We first delimit the scope of our investigation, then state the general research problem, and finally discuss the data and methods used to test specific research hypotheses.

Scope

This paper concerns the flows of staple produce commodities through the internal wholesale marketing system of Cyprus during the year 1970.[1] It is important to keep in mind that this is neither a peasant nor a periodic system, but rather an urban oriented, daily trade mechanism wherein village growers, local dealers, and urban wholesalers have access to well-developed telephone communications and road transport networks through which market price information

---

[1]For the purposes of this analysis, the wholesale marketing process ends when the produce reaches the hands of the retail reseller or institution; or more technically, at the conclusion of the transaction transferring ownership (title) of the commodity from wholesaler to retailer or institution for resale or redistribution to the ultimate consumer.

TABLE 1

PER CAPITA CONSUMPTION OF STAPLE PRODUCE IN CYPRUS 1960-68

(Figures are in okes)[a]

Commodity	Year								
	1960	1961	1962	1963	1964	1965	1966	1967	1968
Tomato	12.7	12.8	13.8	22.9	20.5	21.5	23.9	24.7	25.7
Apple	1.9	1.9	2.6	2.6	4.4	8.6	10.0	13.0	10.3
Cucumber	5.2	4.9	5.2	6.0	5.1	8.7	11.9	15.6	16.1
Potato	33.2	36.6	49.7	47.2	48.3	60.4	44.9	41.4	48.2
Watermelon	16.1	15.3	11.7	22.5	18.9	28.6	25.7	26.9	28.1

Source: Papachristodoulou (1970).

[a]One Cyprus oke = 2.83 pounds avoirdupois = 1.27 kilograms.

and commodities flow at a relatively high degree of efficiency. The structural core of the system consists of a central urban market connected by a radial transport network to an island-wide hinterland. Within the central market at Nicosia, attention is focused on similarities among wholesalers in their transactional links to common sets of supply villages. The direction and magnitude of linkage is measured in terms of "flowvalue," defined as the wholesale value of staple produce shipments from supply origins to wholesaler destinations in the central market. In the hinterland, emphasis is placed on patterns of horticultural land use vis-a-vis the origins of major produce commodity flows. The conceptual approach is thus a blend of commodity flow analysis, and interpretation of flow patterns through a derivation of agricultural location theory.

Over 73 different types of fresh fruits and vegetables are brought to the six municipal wholesale markets of Cyprus in the course of one calendar year. Officials in the Ministry of Finance and in the Department of Statistics and Research suggested at the outset that the scope of our investigation should be limited to those "staple" produce commodities which, by definition, in aggregate account for over one-half the total value of wholesale trade in 1970.

Table 2 lists the total value and percentage figures for the five major produce commodities traded in the municipal wholesale marketing system of Cyprus. Table 3 reveals that the four highest ranking staple produce commodities in terms of wholesale value, also are the top ranking fruit and vegetable commodities in Cypriot dietary expenditure. Most of the staple produce brought to a given municipal wholesale market is consumed in urban households within that municipality.

With the range of our inquiry delimited, we can now lay out a formal statement of the research problem in a format amenable to testing.

Problem Statement

Staple produce commodity flows to the central wholesale market of Cyprus are analyzed to elucidate two key aspects of the internal marketing system: (1) the transactional organization of wholesalers in the marketplace, and (2) the spatial structure of supply in the hinterland.

The general problem consists of two questions around which the analytical framework of Chapters II and III is built:

TABLE 2

WHOLESALE VALUE OF STAPLE PRODUCE BROUGHT TO THE MUNICIPAL WHOLESALE MARKETS OF CYPRUS 1970[1]

Staple Produce Commodity	Total Value	Nicosia	Famagusta	Larnaca	Limassol	Paphos
Tomato	£ 740,321	£ 424,122	£127,636	£ 48,683	£119,953	£19,927
Apple	511,739	268,647	121,767	35,575	67,323	18,427
Cucumber	465,725	274,717	85,135	29,151	63,590	13,132
Potato	226,435	73,550	44,398	19,563	75,015	13,909
Watermelon	189,257	74,551	33,630	15,217	49,142	16,717
Total →	£2,133,477	£1,115,587	£412,566	£148,189	£375,023	£ 82,112
Percent of Total Value (Row)						
Tomato		57.29%	17.24%	06.58%	16.20%	02.69%
Apple		52.50	23.79	06.95	13.16	03.60
Cucumber		58.99	18.64	06.26	13.65	02.82
Potato		32.48	19.61	08.64	33.13	06.14
Watermelon		39.39	17.77	08.04	25.97	08.83
Staple Produce		52.29%	19.34%	06.95%	17.58%	03.85%
Percent of Staple Produce Total (Column)						
Tomato	34.70%	38.02%	30.94%	32.85%	31.99%	24.27%
Apple	23.99	24.08	29.51	24.01	17.95	22.44
Cucumber	21.83	24.63	20.64	19.67	16.96	15.99
Potato	10.61	06.59	10.76	13.20	20.00	16.94
Watermelon	08.87	06.68	08.15	10.27	13.10	20.36

[1]A sixth wholesale produce market is located at Morphou. No summary statistics are available for this market because it is not classified as municipal.

Data from Cyprus (1971c).

TABLE 3

AVERAGE TWO-WEEK PURCHASES OF FRUITS AND VEGETABLES PER URBAN HOUSEHOLD 1971

Selected Produce Purchases		All Fruit and Vegetable Purchases	
Commodity	Average Value (Mils/Household)	Percent of Total Expenditure	Cumulative Percent
Tomatoes	.432	13.35%	13.35%
Potatoes	.288	8.90	22.25
Apples	.263	8.13	30.38
Cucumbers	.195	6.03	36.40
Other Fresh Vegetables	.173	5.35	41.75
Oranges	.169	5.22	46.97
Bananas	.155	4.79	51.76
Haricot Beans, Dry	.140	4.33	56.09
Dry Fruits and Nuts	.132	4.08	60.17
Haricot Beans, Fresh	.093	2.87	63.04
Watermelons	.092	2.84	65.88
	£ 3.236		

Data from Cyprus (1972d).

1. Agora (the marketplace).

Do central market wholesalers form identifiable groups which obtain their produce from common sets of supply villages? In other words, is there a taxonomic grouping of wholesalers based on their flowvalue linkages with common sets of supply origins? This question is largely exploratory in that it deals with the salient features of a little-known relationship which has been broached only tangentially in previous literature.[1] It is hypothesized that

[1]Qualitative typologies of market sellers based on transactional behavior, product specialization, and position in the marketing chain may be found in Trager 1976; V. Smith 1975; Jones 1972; and Vance 1970.

wholesaler destination-supply origin groups do exist, and that their identification can shed some light on a generally obscure aspect of transactional organization in the central market.

2. Promethochora (the supply hinterland).

What are the locational determinants of regional specialization in the central market hinterland? Specifically, how is the level of staple produce flowvalue from supply origins related to land productivity factors and distance to market. The search for the determinants of regional specialization in the market hinterland is guided by the established body of agricultural location theory discussed above, and carries explicit implications for hypothesis testing. It is hypothesized that regional specialization of major produce supply in the hinterland is a simultaneous function of hydrologic, climatic, and market distance variables.

Data

The produce commodity flow data used to test the research hypotheses are unusually accurate and thorough in coverage. This is due primarily to the scrupulous regulation of trade undertaken by the Nicosia municipality.

All Cyprus grown produce destined for retail sale to ultimate consumers in the Nicosia District must, by law, first be channeled through the Nicosia Municipal Wholesale Market. This procedure is similar to England's "Covent Garden System" in which all commodities are sent through one central market at some time in their flow from grower to consumer.[1]

In compliance with Nicosia Municipality Regulations 26, 211, and 212, all fresh produce brought to the Nicosia Municipal Wholesale Market is assessed a gate toll of one-half mil per oke weight, and a market fee of thirty mils per pound wholesale value (The Cyprus Gazette 1968).

To assure proper collection of these municipal taxes on incoming produce flows, unpublished daily trade accounts are meticulously maintained by a crew of municipal clerks working in the marketplace under the direction of Mr. S. Ferekides, Inspector of the Wholesale Markets.

Each wholesale transaction in the marketplace is recorded on a

---

[1]A small volume of fresh fruit and vegetable production marketed in Cyprus does not enter the wholesale sector, being sold at farmgate directly to itinerant retailers, food-processors, and consumers. For an excellent discussion of this process, see Hjiparaskevos (1969).

standardized form listing information on ten variables: (1) wholesaler's name, (2) date, (3) assembly vehicle licence number, (4) village of produce origin, (5) produce type, (6) quantity sold, (7) wholesale price, (8) total value, (9) gate toll, and (10) market fee. The reliability and accuracy of the trade accounts is of paramount importance for they are the sole source authority referred to in adjudicating transaction irregularities, and in the compilation of monthly and annual summaries of produce commodity trade published by the Department of Statistics and Research, Ministry of Finance.

Village of produce origin and wholesaler destination for produce flows to the market are not among the data published in the monthly and annual summaries. Therefore, the complete set of unpublished daily trade accounts covering the normal (i.e., drought-free) calendar year 1970 was borrowed from the Inspector of Wholesale Markets. A team of fifteen Cypriot research assistants edited and codified raw data on the ten variables for 48,902 separate transactions undertaken by a total population of 183 wholesalers during the calendar year 1970.

Next, Mr. P. Mallis, Chief Statistician, Department of Statistics and Research, arranged for the set of codified trade accounts to be keypunched on computer cards. At the University of Chicago Computer Center, the data were copied from cards to magnetic tape, scanned for coding error, and sorted by wholesaler.

Thirteen cases out of the original list of 183 involved entry of both the produce consignor (i.e., the party who delivers produce to an agent for wholesale), and the consignee (i.e., the agent to whom the produce is delivered). All thirteen of these observations were recoded so that the consignee only was listed as the sole wholesaler on the final transaction record. This reduced the total population of wholesalers to 170.

Sample

Based on our discussions with the Inspector of Wholesale Markets, and on our observations of wholesaler behavior in the marketplace, we classified each of the 170 individual wholesalers into one of three mutually exclusive categories, according to mode of transaction. The three general wholesaler types together with their operational definitions are:

1. Merchant Dealers. These full-time professional urban wholesalers receive produce consignments from local dealers, growers, and cooperative agents. Consignments are unloaded at the merchant dealers' rented platform stalls and wholesaled on a commission basis

to retail resellers.

2. Cooperative Organizations. Identified as individual wholesalers on the final transaction inventories, these grower-owned organizations assemble produce from members, and wholesale bulk consignments through merchant dealers to retail resellers.

3. Independent Truckers. This third classification of wholesalers consists of local dealers and private producers. Local dealers in this category assemble produce from village growers for their own account and transport their purchases to the open-air plaza of the Nicosia wholesale market. Growers in the independent trucker category load their crops on hired trucks and carry them to the open-air plaza selling site. As independent truckers, both local dealers and growers park their vehicles in assigned zones in the plaza and tailgate-wholesale their fresh fruits and vegetables directly to retail resellers.

At this point, we therefore had a list of 170 wholesalers which was stratified by transaction mode into three types (i.e., merchant dealer, cooperative organization, and independent trucker).

Next, for each of the 170 cases, total value of trade in pounds Cyprus (i.e., flowvalue) was obtained by summing the wholesale value of each separate transaction in each of the five staple commodities over the entire year.

Wholesalers within each of the three strata were then arrayed in descending rank order by flowvalue. Ninety-four observations with less than £100 flowvalue were purged. This shortened our total count to seventy-six wholesalers.

From a random start, one out of every two cases was drawn from within each of the three rank ordered strata. The result yielded a fifty percent systematic sample consisting of thirty-eight wholesalers engaged in 26,125 separate transactions with 152 supply origins during 1970. A final screening removed sixty-two villages with less than £50 flowvalue, and yielded a 90 x 38 origin-destination flow matrix. Elements of the matrix are value of staple produce flow in Pounds Cyprus, rows are ninety produce supply origins (Table 4, Figure 2), and columns are thirty-eight central market wholesaler destinations (Table 5).

Test Matrices and Methodology

The first research hypothesis is tested by applying principal components analysis to the 90 x 38 flowvalue matrix in order to identify common sets of produce supply origins and wholesaler destinations.

To test the second research hypothesis, a separate hinterland

TABLE 4

PRODUCE SUPPLY VILLAGES AND TOWNS IN ANALYSIS

Case	Census Code	Name
		NICOSIA DISTRICT
1	001	Nicosia Town
2	002	Agrokipia*
3	003	Akaki*
4	007	Alona
5	015	Argaki*
6	016	Argates*
7	018	Astromeritis*
8	019	Avlona*
9	821	Ayia Marina (Khrysokhou) Polis Region, Paphos District *
10	028	Ayios Epiphanics (Soleas)
11	039	Chakistra
12	041	Deftera (Kato & Pano)*
13	053	Galata
14	054	Galini*
15	061	Kakopetria
16	066	Kalopanayiotis
17	070	Kambos
18	075	Kato Kopia*
19	082	Kokkini Trimithia*
20	092	Laxia*
21	103	Mammari*
22	114	Morphou*
23	116	Moutoullas
24	125	Pakhyammos
25	129	Paleometokho*
26	131	Pedhoulas
27	132	Pendayia*
28	133	Pera*
29	138	Pharmakas
30	140	Philia*
31	141	Phlassou
32	145	Politiko*
33	161	Strovolos*
34	163	Tembria
35	178	Zodhia (Kato & Pano)*
		KYRENIA DISTRICT
36	201	Kyrenia Town*
37	215	Kalogrea*
38	230	Lapithos*
		FAMAGUSTA DISTRICT
39	301	Famagusta Town (Varosha)*
40	302	Akanthou*
41	310	Asha*
42	311	Athna (Akhna)*
43	314	Ayia Napa*
44	318	Ayios Evstathios
45	329	Dherinia*

TABLE 4--Continued

Case	Census Code	Name
		FAMAGUSTA DISTRICT--Continued
46	338	Kalopsidha*
47	343	Kondea*
48	352	Liopetri*
49	354	Lysi*
50	367	Paralimni*
51	372	Phrenaros*
52	381	Sinda
53	382	Sotira*
54	393	Vatili*
		LARNACA DISTRICT
55	501	Larnaca Town
56	502	Alaminos*
57	504	Anaphotia*
58	522	Kiti*
59	524	Klavdhia
60	532	Maroni*
61	533	Mazotos*
62	539	Odhou
63	541	Ormidhia*
64	543	Perivolia*
65	548	Pyrga
66	558	Xyolphagou*
		LIMASSOL DISTRICT
67	601	Limassol Town
68	603	Agros
69	608	Kato Amiandos
70	622	Ayios Dhimitrios
71	631	Ayios Yeoryios
72	636	Dhymes
73	642	Kaminaria
74	649	Kilani
75	657	Kyperounda
76	659	Lemithou
77	676	Parekklisha*
78	679	Pelendria
79	683	Phini
80	687	Platres (Kato & Pano)
81	695	Prodhromos
82	696	Pyrgos*
83	703	Trimiklini
84	704	Tris Elies
		PAPHOS DISTRICT
85	801	Paphos Town*
86	811	Argaka*
87	845	Kathikas
88	908	Polis*
89	909	Pomos
90	933	Yialia*

*Origins included in sample of 52 for the test of tomato flow-value determinants.

Fig. 2.--Supply Villages and Towns: Location and Areal Extent

TABLE 5

WHOLESALERS IN ANALYSIS

Case	Name	Classification Criteria
	Merchant Dealers	
1.	Alexandrou, C.	Commission wholesale off enclosed platform stalls rented from the Nicosia Municipality
2.	Alexandrou, M.	
3.	Christou, C.	
4.	Costis & Philiotis	
5.	Hailis, M.	
6.	Hercules & Pantel	
7.	Hji-Georgiou, I.	
8.	Ilias, J.	
9.	Koukkides, R.	
10.	Loizou, C.	
11.	Nikolas, P.	
12.	Paphitis, D.	
13.	Philiotis, C.	
14.	Roditis, C.	
15.	S.E.F.	
16.	Souglides, G.	
17.	Theophanos, P.	
18.	Tsiongouros, M.	
19.	Zittis, A.	
20.	Zitti, E.	
21.	Zodiades, G.	
	Cooperative Organizations	
22.	S.D.G.P. Ayia Marina	Grower-owned institutions which wholesale on consignment through merchant dealers
23.	S.D.G.P. Ayia Napa	
24.	S.D.G.P. Paralimni	
25.	S.D.G.P. Sotira	
26.	Argakas Co-op	
	Independent Truckers	
27.	Dietis, M.	Local dealers and growers who tailgate-wholesale their produce in the open-air plaza, directly to retail resellers
28.	Efstnthiou, S.	
29.	Gelasis, J.	
30.	Georgiou, A.	
31.	Hji-Panteli, J.	
32.	Majoupas, H.	
33.	Nikola, K.	
34.	Petrou, M.	
35.	Samaras, A.	
36.	Shiagolas, A.	
37.	Starris, C.	
38.	Karidis, I.	

attribute matrix was constructed for each of the two highest value produce commodities traded: tomatoes and apples.

Data availability reduced the number of supply origins in the tomato test matrix from the original sample of ninety, to the fifty-two origins marked with an asterisk in Table 4. With these fifty-two supply origins as rows, columns are occupied by one dependent variable (tomato flowvalue), and three predictor variables (hydrology, climate, and market distance).[1] We thus have a 52 x 4 test matrix for the tomato hinterland.

The apple hinterland test matrix consists of the ninety villages in the original sample as rows, and four variables in the columns: one dependent (apple flowvalue), and three independent (hydrology, climate, and market distance). We therefore have a 90 x 4 apple hinterland test matrix.

Mutliple regression analysis is applied to the tomato and apple hinterland matrices to determine the relative influence of environmental and market distance variables on the regional specialization of high value produce supply.[2]

---

[1]Operational definitions of these independent variables are given in Chapter III.

[2]Both the principal components and multiple regression analyses were performed on the IBM 360/370 facility at the University of Chicago Computation Center. See Nie et. al. (1975) for detailed explanations of these catalogued programs.

CHAPTER II

AGORA: TRANSACTIONAL ORGANIZATION OF WHOLESALERS IN THE MARKETPLACE

The first step in attacking the general research problem is directed toward what is perhaps the most vital component in the internal marketing system--the agora. We begin by evaluating Nicosia's situation in the urban and market hierarchies of Cyprus. Site factors relevant to marketplace operation are discussed. Next, the commanding position of urban wholesalers in channeling fresh produce to the central wholesale market is assessed. The introductory groundwork laid, we then address the first research hypothesis by employing principal components analysis to identify groups of central market wholesalers based on the manner in which they obtain produce supplies from common sets of supply areas in the hinterland.

Situation and Site of the Nicosia Municipal Wholesale Market

What is the position of the Nicosia Municipality in relation to its surrounding urban places and municipal wholesale markets? Answers to this question will pave the way to our subsequent examination of marketing channels and flowvalue linkage groups within the Nicosia Municipal Wholesale Market.

Nicosia in the Urban and Market Hierarchies

Nicosia, the administrative, religious, and commercial capital of Cyprus, has dominated the six-town urban hierarchy of the Island since ancient times. In 1970, Nicosia's inhabitants accounted for nearly half of the urban population of Cyprus.

When the six district towns of Cyprus are ranked in descending order of population size, their relationship to each other approaches Zipf's (1941) rank-size rule such that the population of the rth ranking town is roughly $1/r^{th}$ the size of Nicosia. Figure 3 illustrates that a plot of the urban hierarchy on double logarithmic scale is nearly linear or log-normal (Berry and Horton 1970).

In addition to her leadership in the six-town urban hierarchy, Nicosia also tops the five-town wholesale market hierarchy (Table 6). The Nicosia Municipal Wholesale Market handles over half the

TABLE 6

POPULATION SIZE AND WHOLESALE VALUE OF STAPLE PRODUCE TRADE FOR THE FIVE MUNICIPAL WHOLESALE MARKETS OF CYPRUS 1970

Town	Population Size			Wholesale Value of Staple Produce Trade		
	Population Estimate 1970 (000's)	% of Total Town Population	Rank	Wholesale Value 1970 (£ 000's)	% of Total Wholesale Value	Rank
Nicosia	115.0	47.5%	1	1,115.6	52.3%	1
Limassol	51.5	21.3	2	375.0	17.6	3
Famagusta	42.5	17.5	3	412.6	19.3	2
Larnaca	21.4	08.8	4	148.2	07.0	4
Paphos	11.8	04.9	5	82.1	03.8	5
Total	242.2	100%		2,133.5	100%	

Data from Cyprus (1970; 1971c)

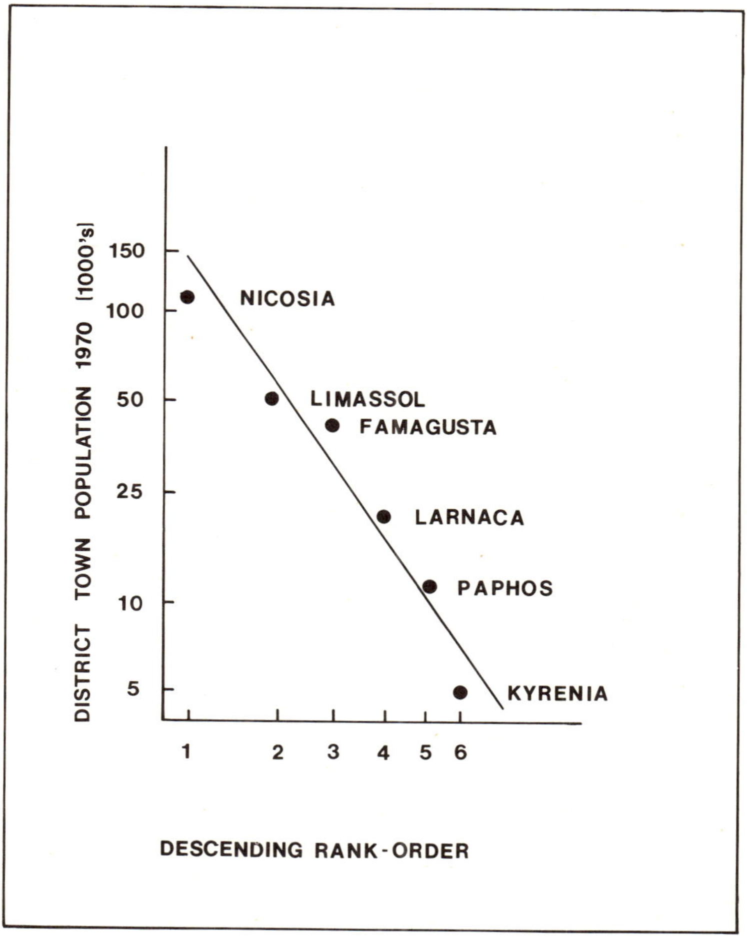

Fig. 3.--Urban Hierarchy. Data from Cyprus (1970).

total value of fruits and vegetables brought to the municipal wholesale markets of Cyprus.

The wholesale produce trade hierarchy plotted on double log scale (Figure 4) approximates the log-normal rank-size distribution of the urban hierarchy. The close parallel between the urban and market rank-size distributions suggests that a well integrated urban wholesale marketing system exists in Cyprus. Inflows of fresh food from the hinterlands appear to be responsive to aggregate urban demand in each of the five district towns which have municipal wholesale markets.

In their study of markets in West Africa, Hodder and Ukwu (1969) note that "for the geographer, one of the most striking features about daily markets is that, unlike periodic markets, they show a clear correlation in their location with the distribution and hierarchy of settlements."

The extent of this correlation in Cyprus is illustrated in Figure 5. This scattergram plot of district town population against wholesale value of staple produce trade indicates a strong positive linear relationship which is confirmed by a Kendall's tau of 0.80 (significant at the 0.05 level of confidence).

Since most of the fresh fruits and vegetables funneled into the municipal wholesale markets are destined for ultimate consumption in urban households, we may conclude that a high degree of association exists between urban consumer demand and the wholesale value of staple produce trade within the five-town market hierarchy of Cyprus.

Figure 6 shows that the Nicosia Municipal Wholesale Market has maintained its leading position within the wholesale produce trade hierarchy over the period 1966-73. Moreover, this semi-logarithmic graph indicates that Nicosia's leadership increased at an increasing rate following the 1968 completion of the new wholesale market facilities on the bypass east of the old walled city. Her rate of change leveled off during the 1971-72 drought, and again accelerated in 1973.

Nicosia's dominance of the urban and market hierarchies is enhanced by her unique situation as the preeminent transport node in the domestic space-economy. Figure 7 illustrates that Nicosia is the hub of an asphalted radial road network which approximates minimum aggregate travel time from the ninety major produce supply origins in our analysis.

Roman (1959), citing Melamid (1958), argues that this road system was developed primarily to serve agriculture, for "the

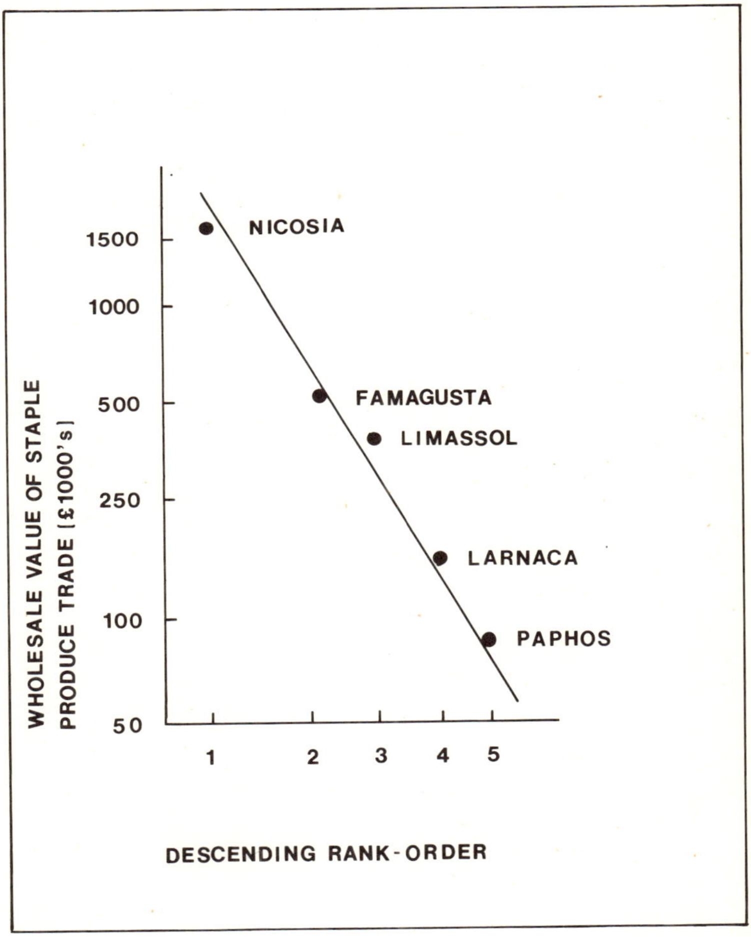

Fig. 4.--Wholesale Produce Trade Hierarchy.
Data from Cyprus (1971c).

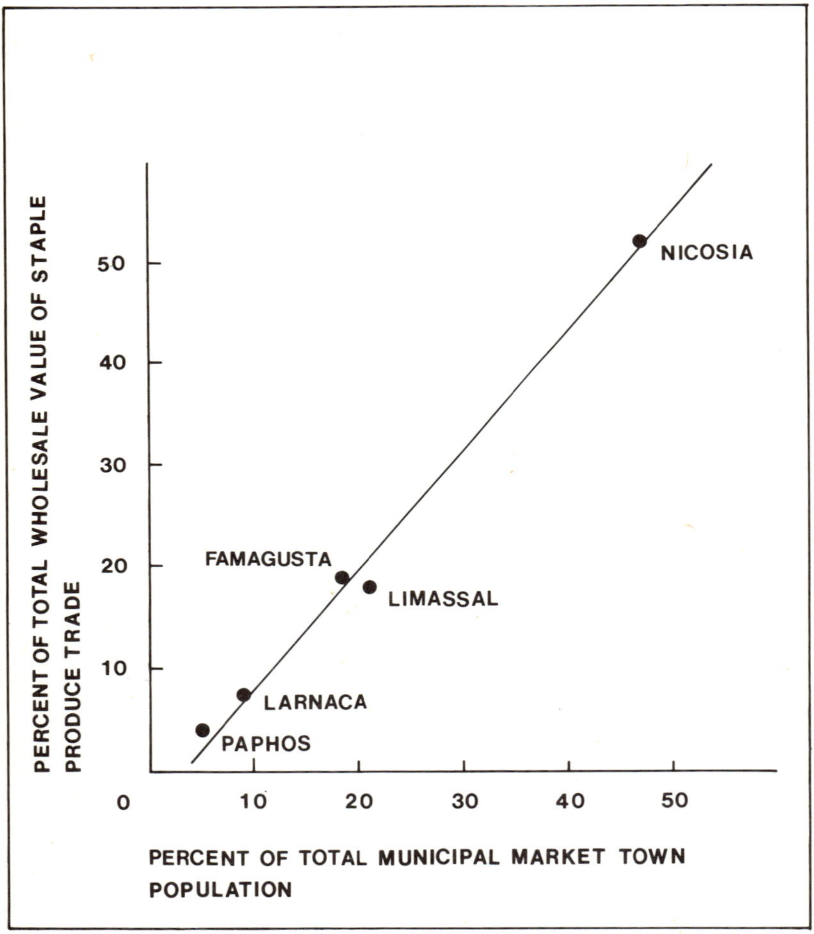

Fig. 5.--Relationship Between Wholesale Produce Trade and Municipal Market Town Population.

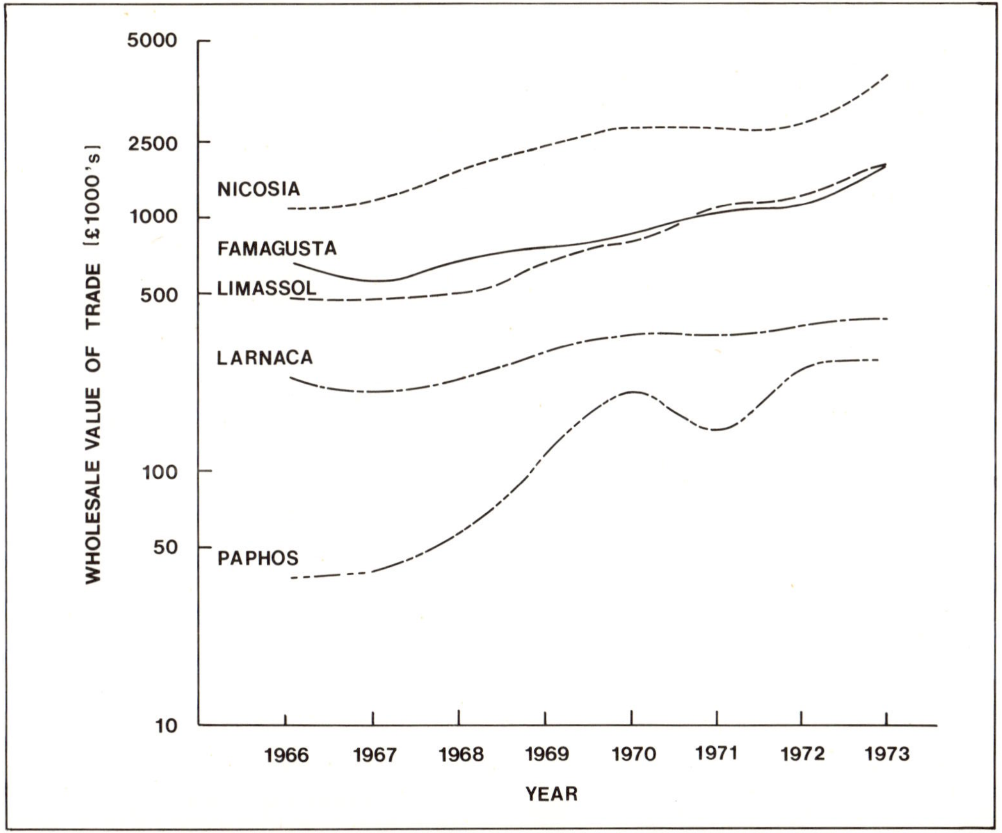

Fig. 6.--Rate of Change in Wholesale Value of Produce Trade for the Five Municipal Wholesale Markets of Cyprus 1966-73.

significance of all types of irrigation to the economy of Cyprus is indicated by the pattern of major roads, which connect practically all irrigated areas." In the light of this observation, the fact that about 96 percent of the vegetable production land is irrigated (Cyprus 1972a) lends further credence to the developmental link between Nicosia's radial road network and horticultural truck farming in the hinterland.

Accessibility to the fresh food supply hinterland was vastly improved by the 1968 relocation of the Nicosia Municipal Wholesale Market from a heavily congested bazaar-like site within the old Venetian walls to the new site two miles to the east (Figure 8). This spacious new facility affords superior accessibility from throughout the Island, and has very good ingress-egress on the two-lane asphalted bypass.

Among the five municipal wholesale markets, only Nicosia has the advantages of uncongested accessibility to a detached, exclusively wholesale facility equipped with modern platform stalls which handle lorry receipts at tail-gate height. These advantages substantially reduce cost of town wholesaling and distribution, and provide better service to produce assemblers, wholesalers, and retail buyers.[1] Each of the smaller town wholesale markets in the hierarchy has higher levels of congestion resulting from their location adjacent to retail facilities in the old town marketplace, and they require lorry-to-ground level unloading.

Nicosia draws the largest value of fruit and vegetable wholesale trade from the most extensive hinterland because consumer demand is highest and transport accessibility easiest. Moreover, shippers are reasonably assured of getting a competitive wholesale price for their entire lorry load in one trip to the central wholesale market.[2] Consequently, larger produce hauls in particular tend to flow past closer but lower level market centers on their way to the central market.

Daily Activity

The functional design of the central market's modern facilities

---

[1]The transition of the Nicosia Municipal Wholesale Market from a "bazaar-like" condition of congested haphazard growth to a modern efficient facility in 1968 parallels the re-organization of the New York City market in 1966 (see Shepherd and Futrell 1970, pp. 478-9).

[2]Personal communication with the cooperative director at the Limassol market.

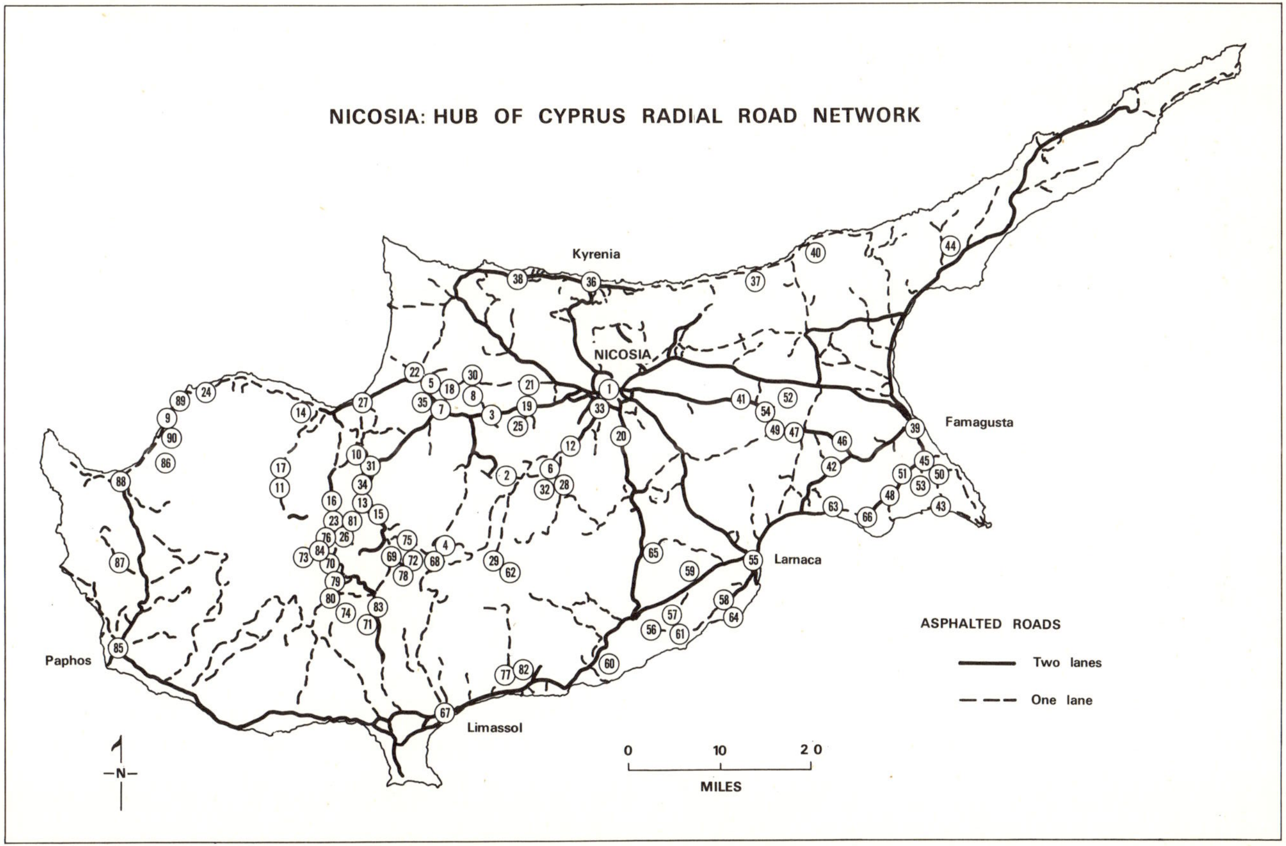

Fig. 7.--Road Network

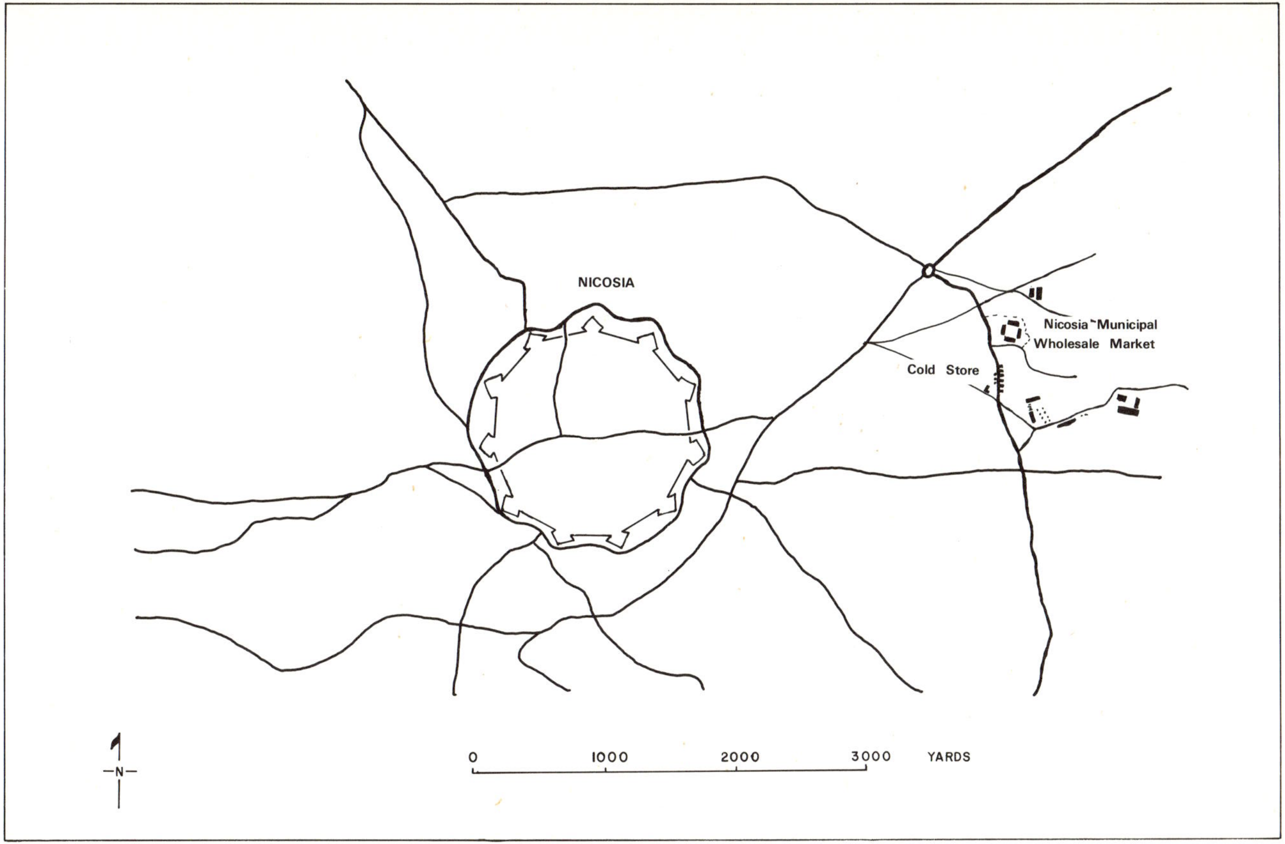

Fig. 8.--Location of the Wholesale Market Within the Nicosia Municipality.

expedites daily trade regulation. Gate tolls of one-half mil per oke produce weight are assessed incoming produce truckers at the main gate (Figure 9). Subsequently, each wholesale transaction is recorded on the standardized form mentioned earlier, and a market fee of thirty mils per pound (£) is levied. Three banks of twelve enclosed platform stalls face inward from the north, east, and south sides of the marketplace (Figure 10). These commission agent selling booths are owned by the Nicosia municipality and are rented annually to merchant dealers. The center of the marketplace is an open-air plaza with assigned parking zones from which independent truckers may tailgate wholesale their produce directly to buyers (Figure 11).

There is a two-phase diurnal cycle of marketplace activity. A period of time is set aside exclusively for produce delivery, running from 10:00 p.m. (12:00 p.m. in winter) to 4:00 a.m. The main gates are then closed for half an hour from 4:00 a.m. to 4:30 a.m., and reopened for the buying period which extends from 4:30 a.m. to 1:00 p.m.

During the delivery phase, growers, local dealers, and cooperative agents hauling fresh produce on vans and lorries enter the main gates where they are weighed-in and hand a "loaded vehicle form" and gate toll to one of the two municipal clerks on tollhouse duty.

Cooperative truckers unload their consignments at merchant dealers' enclosed platform stalls for subsequent commission wholesale to retail resellers during the buying phase. Independent truckers, including growers and local dealers, have the option of either unloading at merchant dealers' stalls for commission wholesale, or of parking their vehicles in the open-air plaza for direct tailgate wholesale to buyers.

The busiest period of wholesale activity occurs during the first two hours of the buying phase. At 4:30 a.m. the main gates are opened to a queue of buyers eager to compare quality and prices, make their purchases (Figure 12), load their vehicles (Figure 13), and get their produce back to their business sites prior to the opening of the retail business day.

The fresh fruits and vegetables are retailed to utlimate consumers in greater Nicosia through several types of outlets. The earliest and largest group of wholesale buyers includes green-grocers who rent stalls at one of the three large municipal retail marketplaces, and neighborhood grocery store merchants. Other wholesale buyers include restauranteurs, food-handling institutions, and street vendors who tailgate retail off pick-up trucks (Figure 14).

Fig. 9.--Main Gates of the Nicosia Municipal Wholesale Market. It is not only the largest, but also the most modern and efficiently run wholesale produce market in Cyprus.

Fig. 10.--North Bank of Enclosed Platform Stalls. Early morning buyers bargain with merchant dealers for the best price, and quickly load their purchases on vehicles for redistribution to retail outlets and institutions throughout greater Nicosia.

Fig. 11.--Tailgating Produce in the Central Plaza. Their open-bed lorry parked in an assigned zone in the open-air plaza, independent truckers adroitly offload watermelon for sale to buyer at center left.

Fig. 12.--Weighing Produce with a Shoulder-beam Scale. Wholesalers supporting the beam scale obtain weight of watermelon in okes. Buyer on far left looks on with a wary eye as the total price of his purchase hangs in the balance.

Fig. 13.--Loading the Day's Supplies. A street vendor loads his early morning purchases on a covered pick-up truck--ready to begin his regular rounds of neighborhood hawking.

Fig. 14.--Haggling with a Neighborhood Hawker. Side-flaps up, and scales at the ready on his tailgate, a street vendor negotiates the price of produce with a Cypriot consumer.

Marketing Channels

Based on the sum of transaction value by wholesaler name given on the final transaction records in the Nicosia Municipal Wholesale Market, incoming produce commodities are handled through three major marketing channels (Table 7). Merchant dealers (channel 1) account for 77.5 percent of total flowvalue; cooperatives (channel 2) handle 19.6 percent; and independent truckers (channel 3) wholesale the remaining 2.9 percent.

Although the growers' cooperative name appears on the final transaction record, cooperative agents always wholesale through merchant dealers in the Nicosia market. Therefore, in terms of who transacts the ultimate transfer of title to retail reseller, merchant dealers may be said to account for 97.1 percent of the total value of trade in the central market. However, for the sake of clarity, in the subsequent analysis of flowvalue linkage groups, the merchant dealer and cooperative channels are treated as two separate entities.

Merchant dealers are commission agents who normally live in Nicosia Town, and rent enclosed platform stalls from the municipality on an annual renewal contract. They wholesale large quantities of crated produce mainly to retail resellers and institutions located throughout greater Nicosia and the Island, as well as to importers from abroad. As the key traders in the central wholesale market, Nicosia merchant dealers set fresh food wholesale prices for the entire internal marketing system of Cyprus.

These urban wholesalers usually undertake to sell produce only on commission, and render their services to local dealers who buy at the village, to cooperative organization agents who assemble members' production at the village, and to growers who bring their produce to them on hired trucks.

Merchant dealers obtain most of their stock through the four-stage marketing chain alternative shown in channel 1 of Table 7. Local dealers who reside in or near the village producing area buy small quantities of produce for their own account from several growers. They then haul their bulked produce by van or lorry to the central wholesale market where they usually consign the shipment to merchant dealers who wholesale it for them on a 12.5 to 20 percent commission basis.

According to Hjiparaskevas' (1969) in-depth study of regional marketing institutions in the south and southwest coastlands of Cyprus, the four-level marketing chain will remain a leading channel of supply for some time, due to the relatively long distance

TABLE 7

MAJOR MARKETING CHANNELS

Major Channel			Marketing Chain Alternatives	
Rank	Percent of Total Value of Trade	Wholesaler on Final Transaction Record	Hinterland Assembly	Market Wholesale
1	77.5%	Merchant Dealer	Grower → Local Dealer →	Merchant Dealer → Retail Reseller
			Grower →	Merchant Dealer → Retail Reseller
2	19.6	Cooperative Organization	Grower → Growers' Cooperative Agent →	Merchant[1] Dealer → Retail Reseller
3	2.9	Independent Trucker	Grower → Local Dealer →	Retail Reseller
			Grower →	Retail Reseller

[1]In channel 2, although the growers' cooperative organization name appears on the final transaction record, cooperative agents always wholesale through merchant dealers in the Nicosia wholesale market (personal communication, Tassos Anastassiades, 1977).

that separates major production areas from the central wholesale market at Nicosia.

In times of scarce supply, the four level chain may often be reduced to the three stage alternative depicted for channel 1 in Table 7. As discussed in detail in the following chapter, early and late season supplies of high-value produce are grown November through April on the warmer frost-free soils near the coast of the more southerly portions of the Island. And, during these months of scarce supply and high prices, Nicosia merchant dealers with commitments to regular customers are sometimes obliged to travel to the villages and buy on their own account, rather than wait for the produce to come to them in town (Hjiparaskevas 1969).

Cooperatives representing the collective interests of many small farmers rank second after merchant dealers in total value of trade. The strength of the cooperative marketing channel is attested to by the fact that the Paralimni and Ayia Napa cooperatives rank third and fourth respectively in total flowvalue among all 38 wholesalers in the sample.

A number of cooperatives have acquired cold storage facilities where fresh produce can be kept for several months. Cold storage helps regulate produce flow into the market so that it is not flooded and prices drop disproportionately during peak normal season production.

Plans are in their final stage to put the cooperative marketing of fresh produce on an Island-wide organized basis with focus on early or out-of-normal season production (Azinas 1973).

There is an appreciable level of seller concentration in wholesale value of trade. Out of the sample of thirty-eight, the four highest value wholesalers together control 47.3 percent of the total value of produce inflow. The first and second highest ranking wholesalers are merchant dealers, and the third and fourth are cooperatives.

In addition, there appears to be a tendency for some of the bigger merchant dealers and cooperatives to obtain produce from higher value supply areas.

The marked degree of seller concentration, the tendency for bigger wholesalers to deal with higher value producing areas, and the growing strength of cooperatives in the marketing chain, together indicate that the Cyprus experience parallels the more advanced stages of marketing system evolution in the models developed by Hodder and Ukwu (1969) and Forman and Riegelhaupt (1970).

Flowvalue Linkage Groups

In the previous section we noted that merchant dealers and cooperative organizations dominate the marketing channels, and observed the tendency for bigger wholesalers to be supplied by higher value producing areas. However, due to the multitude of transactional permutations, the underlying structure of origin-destination flowvalue linkage groups is not readily apparent from inspection of the original 90 x 38 data matrix.

Factor Model

R-mode principal components analysis with varimax rotation is applied to the 90 x 38 flowvalue matrix to reduce the complexity of dyadic linkages and identify common sets of produce supply origins and wholesaler destinations. With reference to our first research hypothesis, the objective is to create an empirical grouping of central market wholesalers on the basis of similar supply origin profiles.

R-mode refers to the cross-sectional 90 x 38 data slice for the year 1970 which is used to delineate patterns of variation among the thirty-eight wholesaler columns in the flow-matrix. The principal components model yields a parsimonious set of wholesaler dimensions which are orthogonal to each other, and are ranked in order of maximal correlation with the original 90 x 38 flowvalue matrix. Factor loadings indicate groups of wholesaler destinations which are alike in the way they assemble produce from supply origins. And, factor scores identify those origins which are major suppliers for each group.[1]

Matrix Transformation

Product-moment correlation coefficients were calculated between each pair of the 38 wholesaler destinations in order to transform the 90 x 38 raw flowvalue to 38 x 38 correlation matrix.

In some cases, the product moment may have been unequally weighted by wholesaler distributions with high positive skew and extreme outliers. However, in this analysis the product moment coefficient is used descriptively rather than inferentially, and therefore a bivariate normal distribution is not assumed

---

[1]SPSS subprogram FACTOR with principal components and varimax rotation options is the specific model applied (Nie et. al. 1975). The fundamental concepts of principal components analysis are well established, and a number of basic references are available on the subject. See, for example, Cattell (1965), Harman (1967), and Rummel (1967; 1970).

(Rummel 1970, pp. 306-7, 366). Moreover, in those runs in which wholesaler distributions were transformed to normality prior to computing the correlation matrix, it was found that the very flowvalue contrasts the coefficients were meant to assess were washed out of the data, resulting in excessively overlapping group factors thereby homogenizing the classification procedure and defeating the original purpose of the analysis. Consequently, it was decided to go with the raw data input which yielded a much clearer grouping of wholesaler destinations.

Factor Interpretation

R-mode principal components analysis, with unities in the diagonal, combined the 38 sample wholesalers into 13 factor groups with eigenvalues greater than 1.0 (Guttman's weakest lower bound). These 13 factors account for 86.9 percent of the variance in the 38 x 38 correlation matrix. In terms of factor complexity, 9 of the 13 factors extracted had eigenvalues less than 2.0, and of these, 3 were group factors with moderate-to-high loadings on pairs of wholesalers, while the remaining 6 were unique factors with one high loading each. Thus, of the nine lower order "groups" of wholesalers, three had shared flowvalue linkages with only two supply villages (factor loadings greater than 0.50), and the remaining six wholesalers had forged their own unique trade links with a single produce supply village each. In short, about two-fifths of the variance in the total value of produce inflows to wholesaler destinations is accounted for by wholesalers who effectively share their source of supply with only one other trader, or they are the sole agent of trade for their particular supply village. Some of these bipolar and unique origin-destination links are of considerable significance in terms of the total value of flow to the market. For example, the village of Paralimni is the third largest supplier of staple produce to the Nicosia Wholesale Market. Of the total of £38,364 originating from Paralimni, £34,630 (90.3%) is wholesaled through the S.D.G.P. Paralimni cooperative organization, and £2,298 (6.0%) is handled by merchant dealer Koukkides. This means that one dominant cooperative wholesaler, and one subordinate merchant dealer together draw over ninety-six percent of the staple produce flows to the central market originating in the third largest supplier in the hinterland. The search for common patterns of flowvalue linkage should not obscure the relative importance of paired and unique wholesaler destinations in the overall inflow pattern.

Application of the Cattell (1966) scree test reduced the 13

initial wholesaler taxa to four orthogonal dimensions with eigenvalues exceeding 2.0. These four major wholesaler groups together account for 56.5 percent of the total flow linkage variance among the 38 individual wholesalers. As an aid to substantive interpretation, varimax orthogonal rotation of the four prinicipal axes simplified the factor structure by maximizing the variance of the squared loadings in each wholesaler group column (Table 8).

There is a descending stepwise ordering of wholesaler groups evident in Table 8, such that those wholesalers with the highest total value of trade tend to cluster together in factor 1 representing the highest percentage of shared variance in produce inflows; and wholesalers with progressively lower rank and size of total flowvalue tend to group together in factors 2, 3, and 4, accounting for respectively declining proportions of shared variance.

Table 9 provides individual identification of each wholesaler destination and supply origin in the four group factors. Each factor may be viewed as a weighted composite of the staple produce inflow patterns of individual wholesaler destinations. The factor loadings in Table 9 identify dominant wholesaler destinations in each group, and the factor scores indicate prominent supply origins. In this way each factor is characterized by a common pattern of produce flow from supply origins to wholesaler destinations in the central market.

Factor 1, accounting for the greatest percentage of total variance (24%) in the correlation matrix, is a clearly delineated group factor with high positive loadings on an intercorrelated cluster of "merchant dealers" (Figure 15). Wholesale value of trade for the twelve constituent merchant dealers ranges from £966 to £50,333; and has a mean of £14,229 (highest of the four groups). In this group are the first (Alexandrou, £50,333), and second (Hji-Georgiou, £37,831) biggest wholesalers from among the sampled thirty-eight.

There are three prominent shared sources of fresh food supply for the factor 1 merchant dealers group. Paphos (£49,863), Ormidhia (£25,455), and Pyrgos (£10,438) rank second, fourth, and eighth respectively in total value of supply to all thirty-eight wholesalers. Eighty-one percent of their combined flowvalue of £85,756 goes to the twelve commission agents in factor 1 (Table 10). Tomatoes and cucumbers are the principal commodities shipped, and Paphos with the highest factor score, sends most (84%) of its production of these two perishables to the merchant dealers group.

Factor 2, "cooperative and merchant dealers," explains fourteen percent of the total variance. This wholesaler group is a

TABLE 8

ORTHOGONAL DIMENSIONS OF WHOLESALER DESTINATIONS
(Varimax Rotated Factor Matrix)[a]

Wholesaler Destination	Principal Loadings				Communality
	Factor 1	Factor 2	Factor 3	Factor 4	
1	0.84				0.77
2	0.93				0.87
3	0.86				0.76
4	0.91				0.85
5	0.67				0.47
6	0.84				0.80
7	0.59				0.56
10	0.67				0.65
11	0.77	0.55			0.89
12		0.94			0.91
15	0.60				0.38
16		0.94			0.89
17		0.85			0.73
18	0.96				0.92
19		0.81			0.87
20	0.52				0.28
21		0.80			0.80
23		0.95			0.91
27			0.65		0.42
29			0.88		0.77
30			0.92		0.86
31			0.99		0.99
32			0.98		0.96
33			0.95		0.90
35				0.97	0.95
36				0.79	0.94
37				0.97	0.95
Eigenvalue	9.12	5.39	4.28	2.66	
Percent of Variance	24.00	14.20	11.30	7.00	
Cumulative Percent	24.00	38.20	49.50	56.50	

[a]Principal axes technique and component analysis model. Factor cutoff at eigenvalue ≥ 2.00. Wholesalers with loadings in the range between plus and minus 0.50 have been omitted to facilitate interpretation.

TABLE 9

WHOLESALER GROUPS BASED ON FLOWVALUE LINKAGES WITH COMMON SETS OF PRODUCE SUPPLY ORIGINS

Wholesaler Destinations		Supply Origins[a]	
Name	Factor Loading	Village or Town	Factor Score
Factor 1: Merchant Dealers			
18 Tsiongouros, M.	0.96	85 Paphos	7.78
2 Alexandrou, M.	0.93	63 Ormidhia	3.79
4 Costis & Philio.	0.92	82 Pyrgos	1.95
3 Christou, C.	0.86		
1 Alexandrou, C.	0.84		
6 Hercules & Pant.	0.84		
11 Nikolas, P.	0.77		
10 Loizou, C.	0.67		
5 Hailis, M.	0.67		
15 S.E.F.	0.60		
7 Hji-Georgiou, I.	0.59		
20 Zitti, E.	0.52		
(24.0% of total variance)			
Factor 2: Cooperative and Merchant Dealers			
23 S.D.G.P. Ayia Napa	0.95	43 Ayia Napa	8.95
12 Paphitis, D.	0.94	63 Ormidhia	1.90
16 Souglides, G.	0.94		
17 Theophanos, P.	0.85		
19 Zittis, A.	0.81		
21 Zodiades, G.	0.80		
11 Nikolas, P.	0.55		
(14.2% of total variance)			
Factor 3: Independent Truckers, Group A			
31 Hji-Panteli, J.	0.99	35 Zodhia	8.21
32 Majoupas, H.	0.98	22 Morphou	4.29
33 Nikola, K.	0.95		
30 Georgiou, A.	0.92		
29 Gelasis, J.	0.88		
27 Dietis, M.	0.65		
(11.3% of total variance)			
Factor 4: Independent Truckers, Group B			
35 Samaras, A.	0.97	1 Nicosia	9.09
37 Starris, C.	0.97		
36 Shiagolas, A.	0.79		
(7.0% of total variance)			

[a]Only those origins with orthogonal factor scores $\geq$ 1.50 are included.

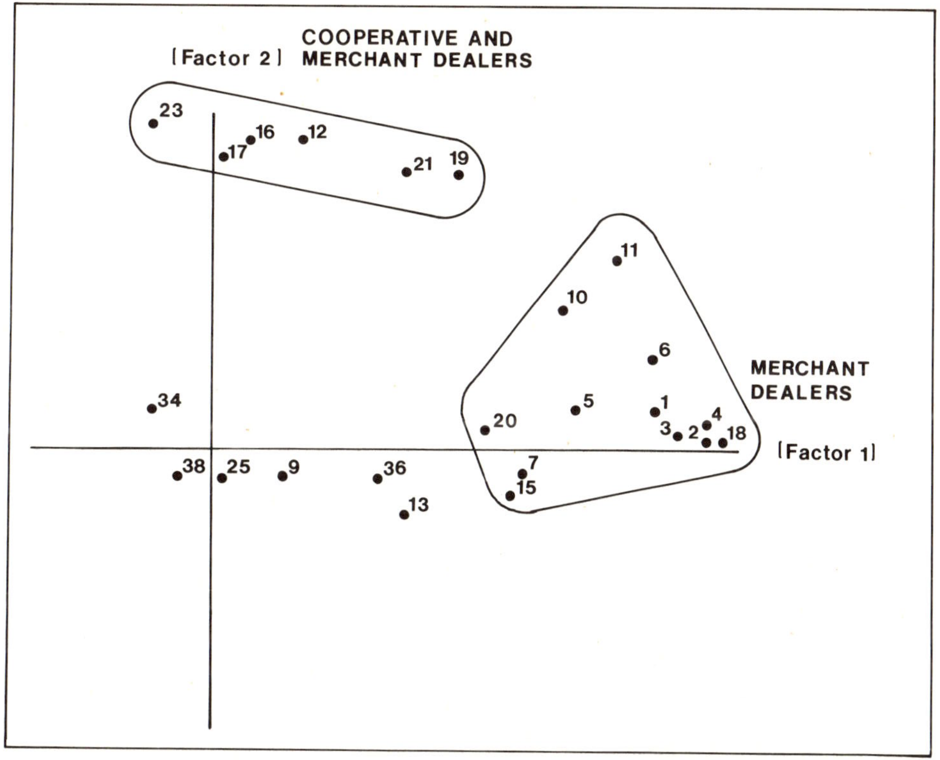

Fig. 15.--Plot of Factor Score Coefficients for Merchant Dealers (Factor 1), and Cooperative and Merchant Dealers (Factor 2).

TABLE 10

WHOLESALER GROUP AND COMMON SUPPLY ORIGIN FLOWVALUE (In Pounds Cyprus)

Wholesaler Group			Common Supply Origin							
				Staple Produce		Specific Commodity				
Factor Number	Name	Flowvalue From All 90 Origins	Name	To Factor Group	To All 38 Wholesalers	Tomato	Apple	Cucumber	Potato	Water-melon
1	Merchant Dealers	£170,748	Paphos	£41,645	£ 49,863	£17,032	£184	£28,287	£ 776	£3,584
			Ormidhia	19,214	25,455	22,821	27	2,075	507	25
			Pyrgos	9,015	10,438	10,154	80	101	27	76
			Total	£69,874	£ 85,756	£50,007	£291	£30,463	£1,310	£3,685
2	Cooperative & Merchant Dealers	£ 83,679	Ayia Napa	52,318	55,973	44,270	25	10,334	147	1,197
			Ormidhia	6,241	25,455	22,821	27	2,075	507	25
			Total	£59,559	£ 81,428	£66,091	£ 52	£12,409	£ 654	£1,222
3	Independent Truckers (A)	£ 3,470	Zodhia	2,020	3,611	1	42	...	390	3,178
			Morphou	909	2,247	217	1	58	17	1,954
			Total	£ 2,929	£ 5,858	£ 218	£ 43	£ 58	£ 407	£5,132
4	Independent Truckers (B)	£ 4,580	Nicosia[1]	£ 1,829	£ 7,295	£ 229	£4,858	£ 75	£ 369	£1,764

[1]Cold storage commodities originate in Nicosia

combination of one large growers' cooperative, and six intermediate value wholesalers (Figure 15). Value of trade for these seven wholesalers is high to intermediate, ranging from £1,240 to £24,120. Mean value is 11,954. Growers' Cooperative S.D.G.P. Ayia Napa with a total value of trade from all ninety origins of £24,120, is the fourth largest wholesaler, and the second ranking cooperative in our sample.

Of the two common supply origins for factor 2, Ayia Napa village has the highest factor score, and is the single biggest (£55,973) total value producing area out of the ninety origins in our sample. This dominant supplier sends practically all (95%) of its primarily tomato and cucumber crops to the factor 2 group of wholesalers (Table 10).

Factor 3, "independent truckers, group A" (11% of total variance) provides a distinctively different group of wholesalers in comparison to factors 1 and 2. (Figure 16). It is composed exclusively of six private growers and local dealers who tailgate wholesale directly to buyers in the open air plaza of the central market. This group of independent truckers is in the small trade range (range = £187 to £1,751; $\bar{x}$ = £578).

The two common supply villages, Zodhia and Morphou, specialize in shipping watermelons to the Nicosia wholesale market. Zodhia, the leading supply origin, wholesales about fifty-six percent of its watermelons through independent truckers, group A.

Factor 4, "independent truckers, group B" (7% of total flow-value) is made up of three local dealers who are at the intermediate to small level on the trade scale (range = £70 to £4,161; $\bar{x}$ = £1,527). Figure 17 illustrates the plot of factor score coefficients for this group of independent truckers who trade mainly in apples and watermelons originating in cold storage facilities located near the Nicosia Municipal Wholesale Market.

Summarizing these results, we find that nearly three-fifths of the total variance in the value of fresh produce inflow is accounted for by four groups of wholesalers--each of which shares its supply from a common set of origins.

As we might expect from what has been said earlier, the merchant dealers' and cooperative organizations' dominance of the marketing channels is reflected in the composition of factors 1 and 2 which explain twenty-four and fourteen percent of the total variance respectively. Moreover, within these two groups, the tendency of higher value merchant dealers and cooperatives to share flow-value linkages with higher value supply areas is again apparent.

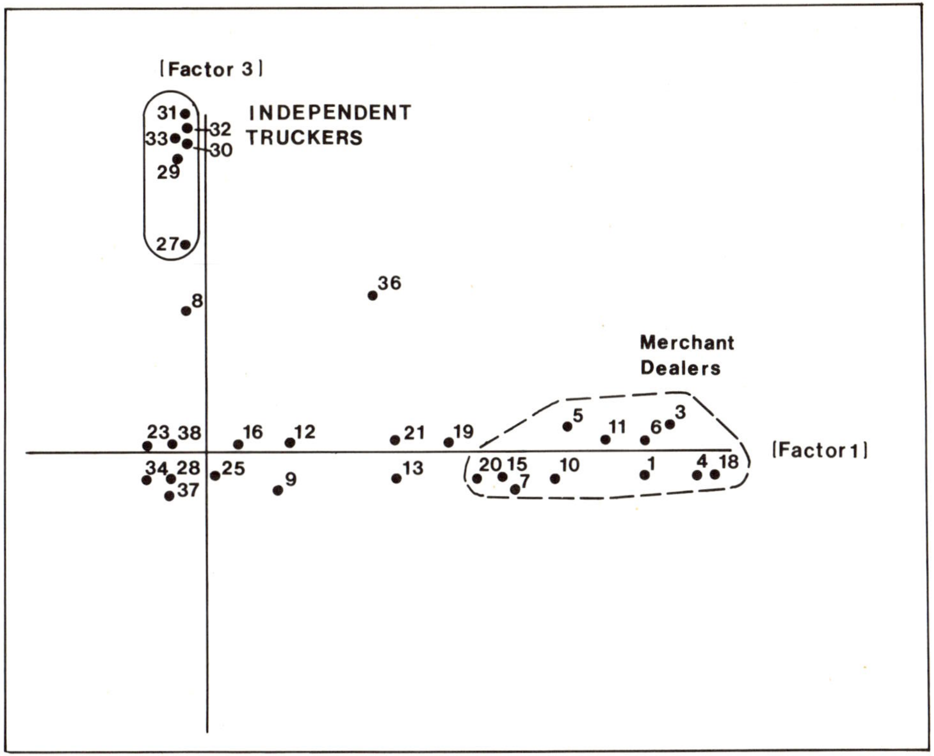

Fig. 16.--Plot of Factor Score Coefficients for Independent Truckers, Group A (Factor 3).

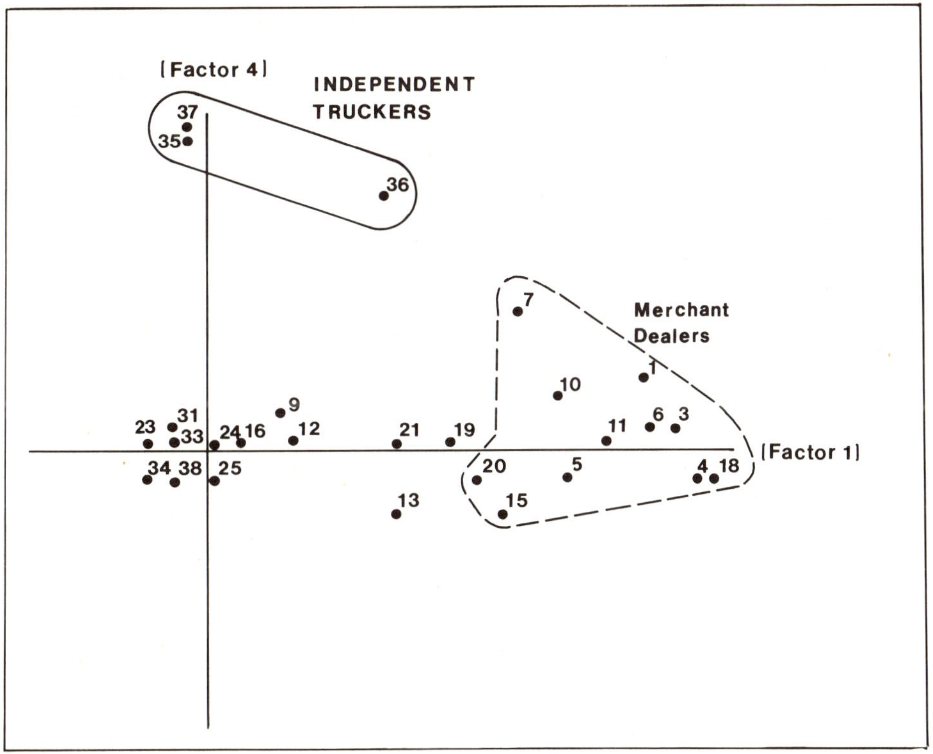

Fig. 17.--Plot of Factor Score Coefficients for Independent Truckers, Group B (Factor 4).

Also important is the finding that two-fifths of the total variance in produce flowvalue linkages is accounted for by paired and unique wholesaler destinations which obtain staple fruits and vegetables from only one supply origin. Many of these narrowly defined trade connections are between smaller value producers and wholesalers.

However, the presence of the larger growers' cooperatives is clearly felt in both the paired and the multiple wholesaler destination groups. Ninety percent of Paralimni's village production is wholesaled through the Paralimni Growers' Cooperative. And, forty-three percent of Ayia Napa's village production is wholesaled through the Ayia Napa Growers' Cooperative. These two examples attest to the growing significance of the cooperative movement discussed by Azinas (1973). There appears to be a trend toward cooperative monopolization of both the supply and wholesale disposal of fresh food in Cyprus. These growers' cooperatives enable small-scale produce farmers to benefit commercially from large-scale economies in assembly, transport, and wholesale. This is one avenue for closing the development gap between increased production among small landholders and growing demand for fresh food in urban market centers.

In this chapter we have examined the transactional organization of wholesalers in terms of marketplace activity, marketing channels, and flowvalue linkage groups. In the following chapter, these central market wholesalers are treated in aggregate, as a single destination, as we turn our attention to the spatial structure of produce supply in the hinterland.

CHAPTER III

PROMETHOCHORA: SPATIAL STRUCTURE OF PRODUCE SUPPLY IN THE HINTERLAND

Two related questions are dealt with in this chapter: where are the staple produce supply areas for the central wholesale market located, and its corollary, why are they located where they are? In answer to these questions, first the salient patterns of staple produce flows from multiple hinterland origins to the central wholesale market destination are depicted in a series of six flow maps. Next, the process behind the dominant patterns are described with reference to land productivity resources and accessibility to market. Finally, a multiple regression model of horticultural land use is tested using tomato and apple flowvalues to the Nicosia Wholesale Market.

The Pattern of Staple Produce Flows

Wholesale value of staple produce commodities shipped by lorry from major supply villages to the Nicosia Wholesale Market in 1970 are shown in Figures 18 through 23.[1] Origin-destination flow patterns for tomato, apple, cucumber, potato, watermelon and aggregate flows are mapped by centering on each major supply village a circle whose radius is proportionate to the value of flow to the central market, and then linking each village origin to the market.

Figure 23, the composite major produce flow map, summarizes the salient spatial pattern running through the set of five individual flow maps. The outstanding theme is that the value of fresh food supply increases with increasing distance outward from the central market.

The fifteen village concentric truck-farming belt shown in Figure 23 supplies over four-fifths (81.1%) of the staple produce commodity flows to the Nicosia Municipal Wholesale Market. This semicircle of supply is located at a mean radial road distance of 51

---

[1]Major supply villages are those accounting for greater than one percent of the total flowvalue from all 90 origins in the sample

TABLE 11

PRODUCE COMMODITY FLOWS TO THE NICOSIA MUNICIPAL WHOLESALE MARKET BY ORIGIN[a]

Case Number	Origin	Census Code Number	Flowvalue (Pounds Cyprus)	% of Total Flowvalue	Cumulative %
			Tomato		
43	Ayia Napa	314	£ 44270	28.83	28.83
50	Paralimni	367	29552	19.24	48.07
63	Ormidhia	541	22821	14.86	62.93
85	Paphos	801	17032	11.09	74.02
77	Parekklisha	676	12865	8.38	82.40
82	Pyrgos	696	10154	6.61	89.01
29	Pharmakas	138	2630	1.71	90.72
49	Lysi	354	2621	1.71	92.43
	Total 90 Origins		£ 153570		
			Cucumber		
85	Paphos	801	£ 28287	39.06	39.06
43	Ayia Napa	314	10334	14.27	53.33
47	Kondea	343	9412	13.00	66.33
50	Paralimni	367	4601	6.35	72.68
49	Lysi	354	3700	5.11	77.79
40	Akanthou	302	2755	3.80	81.59
86	Argaka	811	2099	2.90	84.49
63	Ormidhia	541	2075	2.86	87.35
19	Kokkini Trimithia	82	1523	2.10	89.45
9	Ayia Marina (Khrysokhou)	821	938	1.30	90.75
37	Kalogrea	215	860	1.19	91.94
21	Mammari	103	856	1.18	93.12
41	Asha	310	775	1.07	94.19
	Total 90 Origins		£ 72423		
			Apple[b]		
74	Kilani	649	£ 11172	26.27	26.27
80	Platres, Kato	687	6534	15.36	41.63
81	Prodromos	695	4145	9.75	51.38
10	Ayios Epiphanos	28	3814	8.97	60.35
79	Phini	683	3022	7.11	67.46
34	Tembria	163	1863	4.38	71.84
72	Dhymes	636	1386	3.26	75.10
4	Alona	7	1257	2.96	78.06
15	Kakopetria	61	1173	2.76	80.81
70	Ayios Dhimitrios	622	1168	2.75	83.56
69	Kato Amaindos	608	1018	2.39	85.95
13	Galata	53	964	2.27	88.22
23	Moutoullas	116	796	1.87	90.09
26	Pedhoulas	131	587	1.38	91.47
78	Pelendria	679	571	1.34	92.81
75	Kyperounda	657	502	1.18	93.99
11	Chakistra	39	457	1.07	95.06
	Total 90 Origins		£ 42526		

TABLE 11--Continued

Case Number	Origin	Census Code Number	Flowvalue (Pounds Cyprus)	% of Total Flowvalue	Cumulative %
		Watermelon			
85	Paphos	801	£ 3584	20.05	20.05
50	Paralimni	367	3271	18.29	38.34
35	Zodhia, Kato	178	3178	17.78	56.12
22	Morphou	114	1954	10.92	67.04
43	Ayia Napa	314	1197	6.70	73.74
45	Dherina	329	1180	6.60	80.34
53	Sotira	382	684	3.83	84.17
86	Argaka	811	671	3.75	87.92
51	Phrenaros	372	346	1.94	89.86
	Total 90 Origins		£17878		
		Potato			
48	Liopetri	352	£ 2657	14.50	14.50
53	Sotira	382	2546	13.90	28.40
58	Kiti	522	2252	12.29	40.69
42	Athna	311	1847	10.08	50.77
64	Perivolia	543	1317	7.19	57.96
50	Paralimni	367	857	4.68	62.64
41	Asha	310	844	4.61	67.25
85	Paphos	801	776	4.24	71.49
47	Kondea	343	694	3.79	75.28
19	Kokkini Trimithia	82	520	2.84	78.12
63	Ormidhia	541	507	2.77	80.89
39	Famagusta	301	442	2.41	83.30
35	Zodhia, Kato	178	390	2.13	85.43
1	Nicosia	1	369	2.01	87.44
18	Kato Kopia	75	270	1.47	88.91
46	Kalopsidha	338	264	1.44	90.35
	Total 90 Origins		£18321		
		Major Produce			
43	Ayia Napa	314	£ 55973	18.37	18.37
85	Paphos	801	49863	16.36	34.73
50	Paralimni	367	38281	12.57	47.30
63	Ormidhia	541	25455	8.35	55.65
77	Parekklisha	676	12914	4.24	59.89
47	Kondea	343	11378	3.73	63.62
74	Kilani	649	11172	3.67	67.29
82	Pyrgos	696	10438	3.43	70.72
80	Platres, Kato	687	6534	2.14	72.86
49	Lysi	354	6508	2.14	75.00
81	Prodhromos	695	4177	1.37	76.37
53	Sotira	382	3926	1.29	77.66
10	Ayios Epiphanios	28	3850	1.26	78.92
35	Zodhia, Kato	178	3611	1.19	80.11
86	Argaka	811	3169	1.04	81.15
	Total 90 Origins		£304718		

[a]Village and town administrative areas which supply ≥ 1% of staple produce flowvalue are listed here. Original sample consists of 90 origins

[b]Apple figures exclude 4,585 originating in Nicosia cold storage facilities.

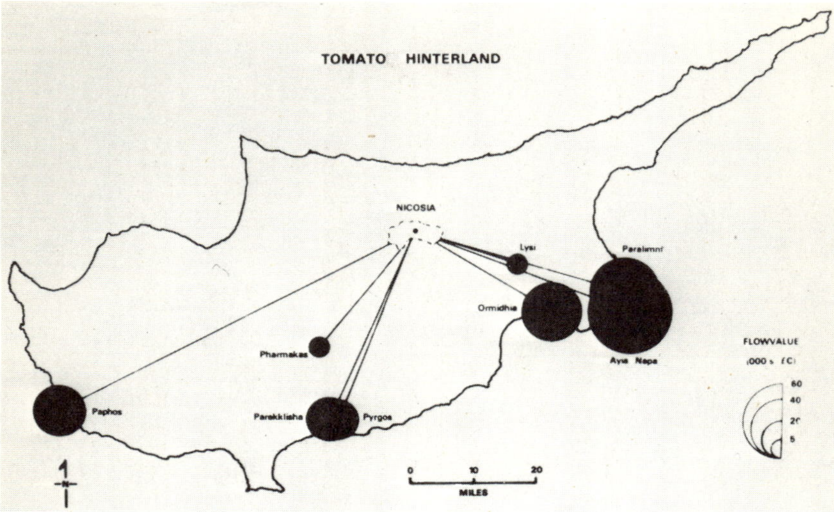

Fig. 18.--Tomato Hinterland

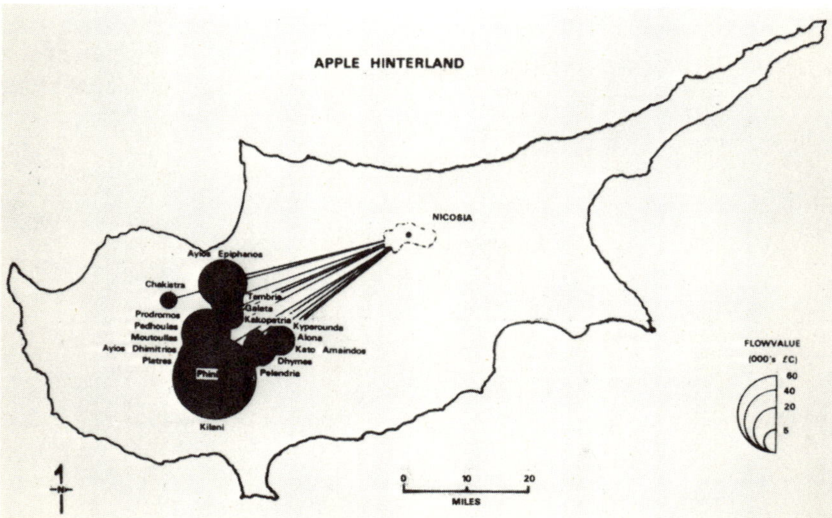

Fig. 19.--Apple Hinterland

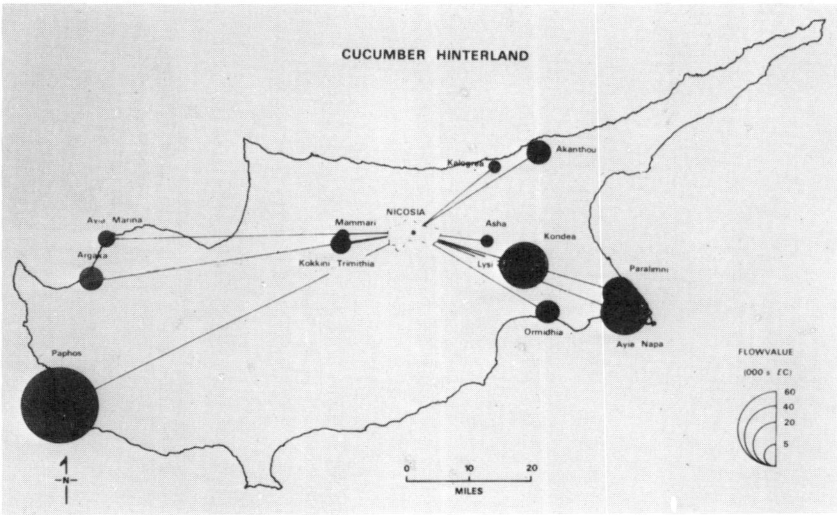

Fig. 20.--Cucumber Hinterland

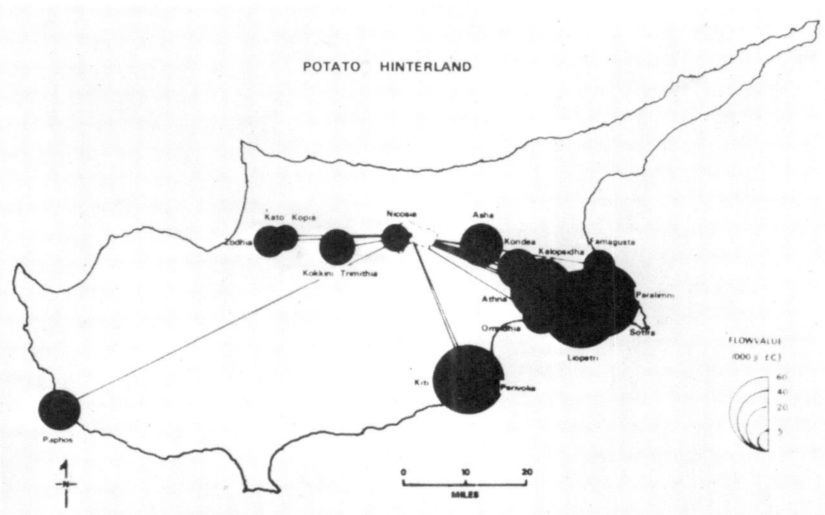

Fig. 21.--Potato Hinterland

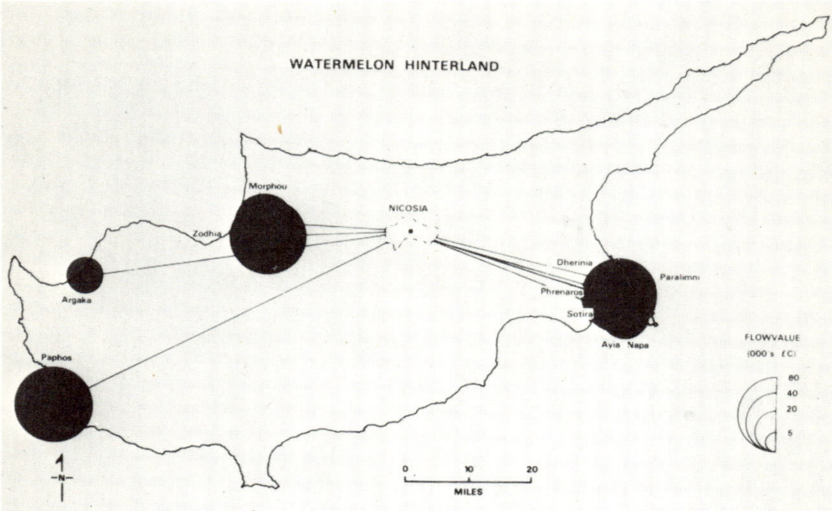

Fig. 22.--Watermelon Hinterland

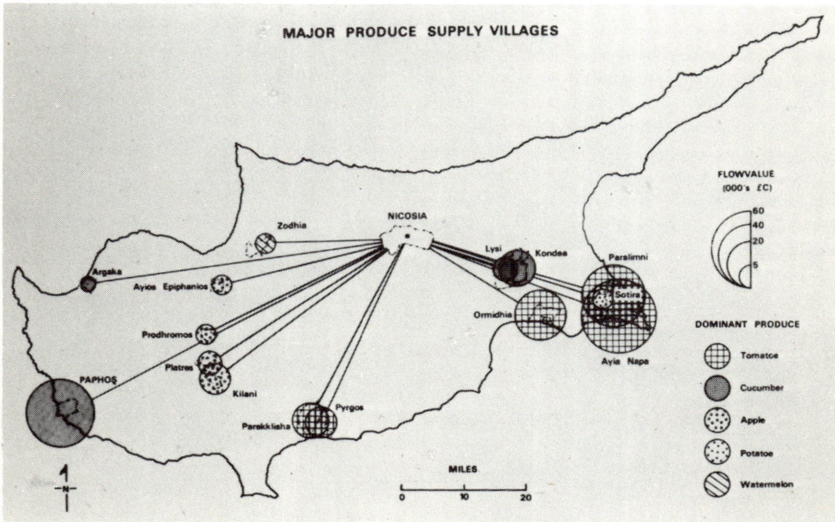

Fig. 23.--Major Produce Supply Areas

miles south of the central market.

Within the specialized truck-farming belt, five high-value supply areas together account for three-fifths (59.9%) of major produce flowvalue: Ayia Napa (18.4%), Paphos (16.4%), Paralimni (12.6%), Ormidhia (8.3%), and Parekklisha (4.2%). These five dominant supply origins, at a mean distance of 56 miles from Nicosia, are distributed along an arc of aquiferous lowlands bordering the Island's southern seacoast.

The highest value supplier of staple produce, Ayia Napa, is situated on the rolling and irregular plains of the Kokkinchoria Peninsula--51 miles southeast of the Nicosia market. And, the second ranking supplier, Paphos, is located along the southwest coastal lowlands at a shortest route distance of 94 miles from Nicosia.

Assuming favorable driving conditions, and an average speed of 50 miles per hour, it would take a trucker hauling fresh produce from Paphos one hour and fifty-three minutes to reach the Nicosia market. In addition to the nearly two hour journey to market, actual transport cost to the trucker would range between £1.880 and £2.820 per ton.[1]

Figure 24 illustrates the weak positive relationship between staple produce flowvalue and distance to market. Kendall's tau for the fifteen leading supply origins ($\tau = 0.18$) is equivalent to Pearson's correlation coefficient for all ninety origins ($r = 0.18$). Both measures of association are significant at the 0.05 level of confidence. This result is the reverse of what would be expected using the gravity model approach to produce commodity flows (Harris and Carrothers 1956).

The pattern of increasing flowvalue with increasing distance from the central market represents a radical departure from the pre-1930's Thünen pattern of high intensity fruit and vegetable production concentrated in market gardens immediately adjacent to Nicosia, and within a maximum radius of ten miles from town--about one hour's journey by donkey-track (Christodoulou 1959). Moreover, these findings are in direct contrast to the spatial arrangement of vegetable supply areas found in Australia (D. Smith 1966), New Zealand (Hunt 1959), and Japan (Eyre 1959). In each of these cases, the intensity of production and the proportion of perishability

---

[1]Actual transport cost is estimated at between £0.020 and £0.030 per ton-mile in 1970 (Tassos Anastassiades 1977: personal communication).

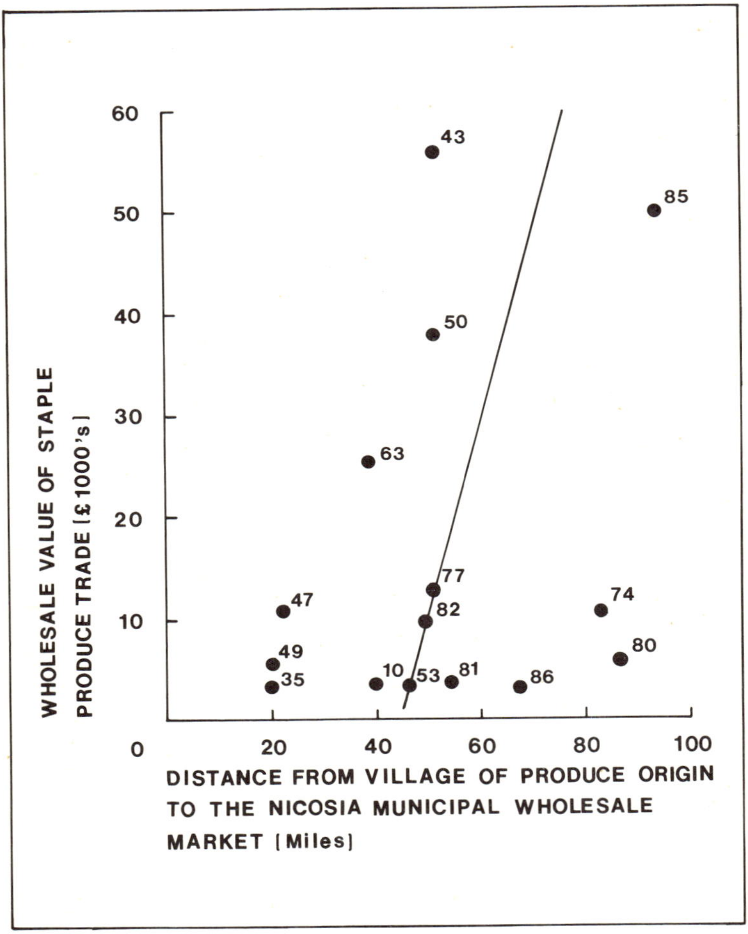

Fig. 24.--Relationship between Wholesale Value of Staple Produce Trade and Distance from Village of Produce Origin to the Nicosia Municipal Wholesale Market. Figures identify fifteen villages whose flowvalue exceeds one percent of the total value of trade for the ninety origins in the sample.

decreases outward from the market. The contrast is perhaps most poignant in juxtaposition to Eyre's observation that 80 percent of Tokyo's fresh vegetable supply "originates in a remarkable truck-gardening zone that lies on all sides of the advancing city front ..." its best developed and most productive parts are concentrated within a 30-mile radius of the center of Tokyo (1959, p. 458, italics mine). He concludes

> The most striking fact concerning the sources of Tokyo's fresh-food supply. . .is that so much of the food originates within a short distance of the city (474).

Contrapuntally, to paraphrase Eyre, it may be said that the most striking fact concerning the sources of Nicosia's fresh-food supply is that so much of the food originates at such a (relatively) long distance from the city.

On an aggregate commodity flow level, there is a remarkable similarity between the United States and Cyprus in both the pattern and process of specialized truck-farming. High value supply areas are located along the frost-free southern extremities of the land mass, near the meliorating influence of the sea, and on irrigated lowlands. It is in these early-season localities that the hydro-climatic advantages outweigh the cost of shipping produce to markets in the north.

Thus far, our attention has been focused on the arrangement of staple produce commodity flows on an aggregate level. Next, we turn to an investigation of specific patterns of regional specialization for tomato and apple supply hinterlands.

Regional Specialization of Tomato and Apple Hinterlands

Tomato and apple commodity flows are selected for detailed analysis because they are the two leading value of flow commodities in the central wholesale marketing system of Cyprus, together accounting for over 60 percent of Nicosia's value of trade in staple produce during 1970. Secondly, almost the entire production of tomatoes and apples in Cyprus is consumed in fresh form in the domestic market (Papachristodoulou 1970). And, third, tomato and apple supply villages form two distinct nearly-homogeneous regions of horticultural specialization in Cyprus. Tomatoes, represent lowland borehole and well-irrigated vegetables; and apples, the highland spring and stream irrigated deciduous tree fruits.

Comparison of tomato and apple hinterland distributions (Figures 18 and 19) with the spatial arrangement of elevation, precipitation, and infiltration (Figures 25, 26, and 27) suggests the following dichotomy of regional specialization.

Aquiferous Lowland Tomato Region

Tomatoes, the highest value, lowest bulk, and most perishable of the major produce commodities, are grown in greatest proportion furthest from the central wholesale market, and nearest to the sea-coast. Three "warm soil" nodes of high value tomato supply are distributed along the southern frost-free coastlands: (1) the Kokkinochoria Peninsula, including the villages of Ayia Napa, Paralimni, and Ormidhia, (2) the village of Parekklisha, and (3) the Paphos area. These three origins not only provide 60 percent of total staple produce flowvalue, but also supply over 82 percent of the total value of tomato shipments to Nicosia.

All three supply nodes average less than 400 feet above mean sea level. They specialize in growing low-temperature varieties of tomatoes in sheltered pockets along the southern coastlands with the help of low polyethylene tunnels (Baldini 1968).

The Kokkinochoria Peninsula is the leading tomato supply area, sending 63 percent of the total value of tomato flows to the Nicosia market. The early season production of tomatoes in the Kokkinochoria Peninsula is made possible by the same combination of climatic and hydrologic advantages which facilitate early tomato production in the Florida peninsula of the United State (Jumper 1969; Pierce 1975).

Southern exposure combined with proximity to warm sea waters moderate climatic extremes, providing a long frost-free growing season with mild winter temperatures, and a moderate range in annual and diurnal nyctotemperature, associated with high relative humidity--all providing an especially favorable micro-climate for the cultivation of tomato plants (Went 1957). Christoudoulou (1959, p. 148) makes the following observation regarding the favorable situation of the Kokkinchoria Peninsula:

> Paralimni with its peninsular position in the southeast of Cyprus and bathed by the warmest sea of Cyprus is the centre of early vegetables _par excellence_. Furthermore, its gardens are on the sandy raised beach and protected by the ridge of sand-dunes from inland cold winds.

In addition to these climatic advantages, the area is underlain by a first-class sand aquifers which supply almost 99 percent of the irrigation water (Figure 28). Most of the tomato truck gardens are irrigated by water mechanically pumped from the sand aquifers

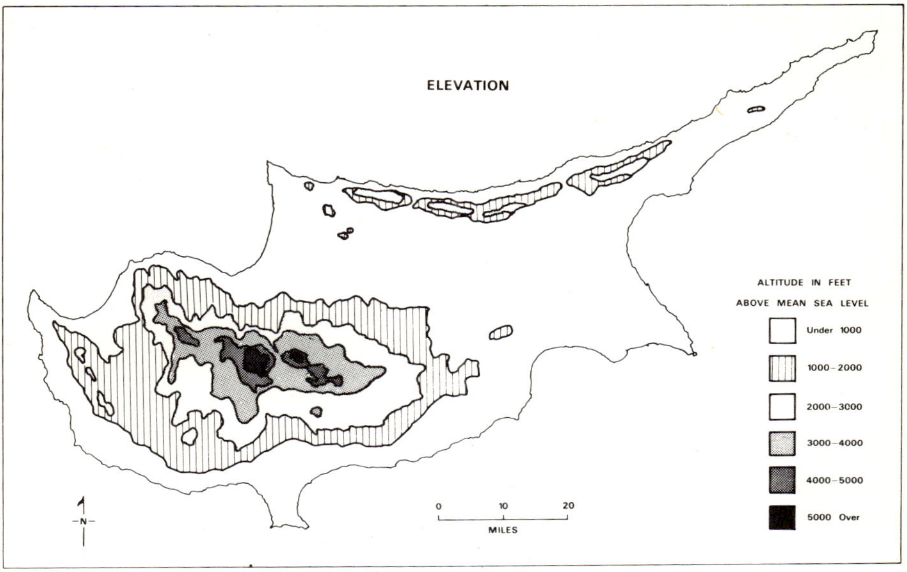

Fig. 25.--Elevation

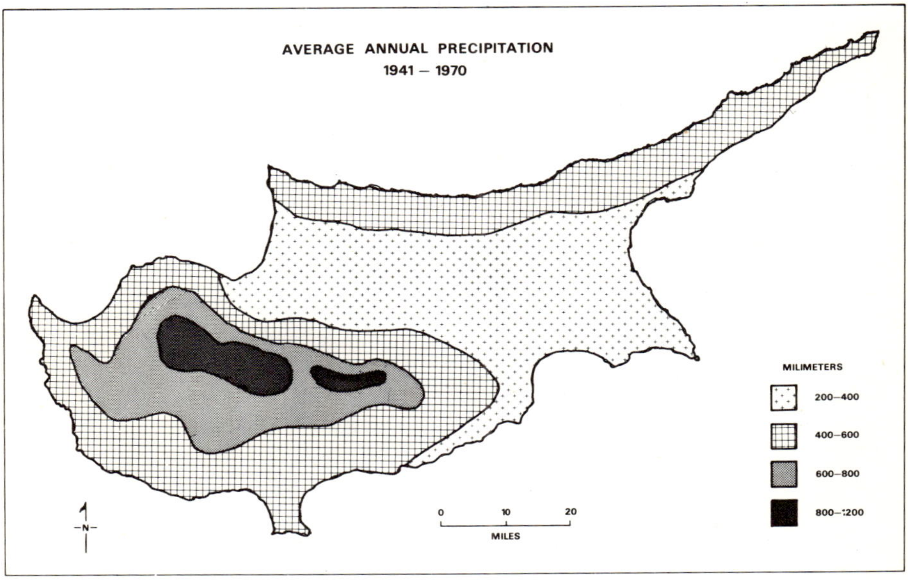

Fig. 26.--Average Annual Precipitation 1941-70

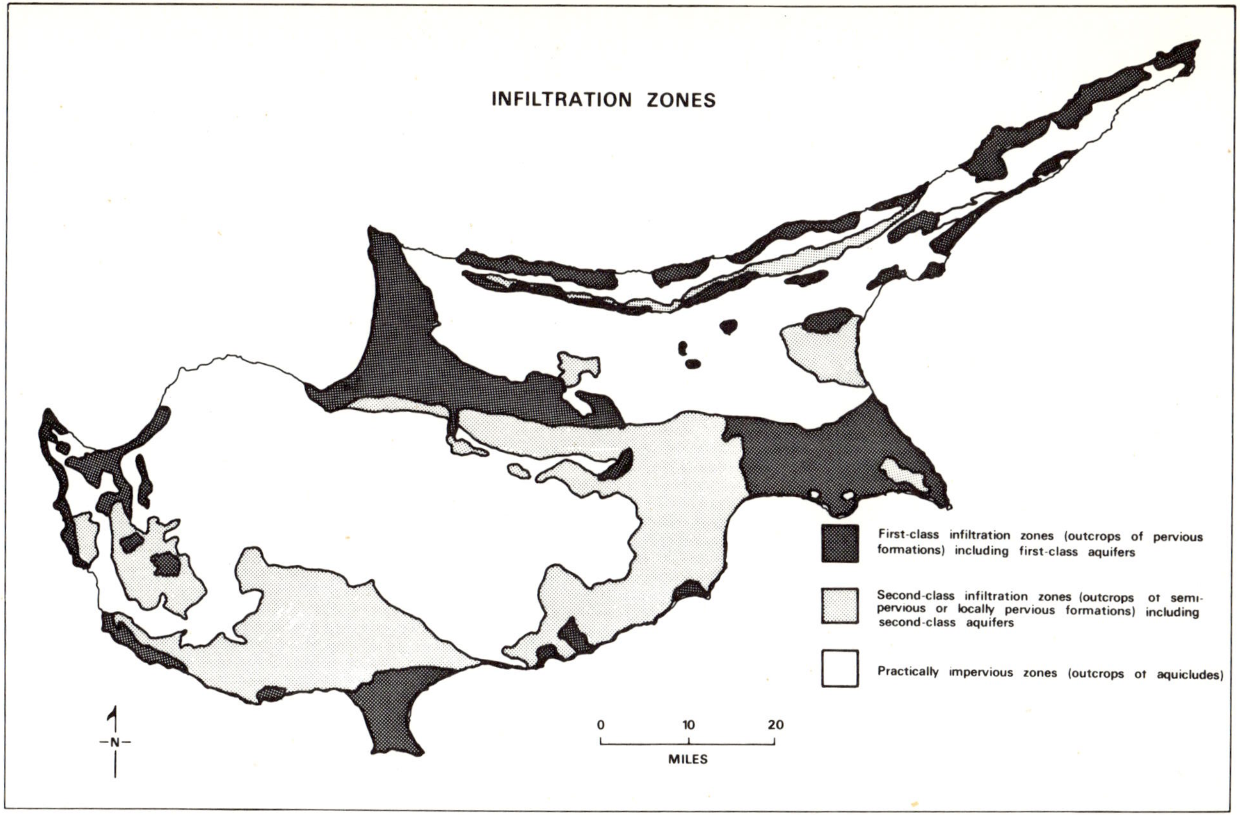

Fig. 27.--Infiltration Zones

through steel casing boreholes, and wells.[1]

The under-cover cultivation of early and late season tomato crops commands high prices in the Nicosia market. The labor-intensive harvest period for tomatoes grown under cover (Table 12) corresponds with the peak early season wholesale price period extending from March through May in the Nicosia Wholesale Market (Figure 30). The wholesale price drops May through August as normal (open-air) summer season crops flood the market, and rises August through November during the late season period of short supply.

Rainfed Highland Apple Region

Major apple supply villages are rather tightly clustered inland in the rainfed cooler upper valleys of the Troodes Massif (Figures 19, 25, and 26). Highest flowvalues originate in the mountain villages of Kilani (2625'), Platres (3281'), and Prodhromos (4330'). These three mountain villages together supply 51 percent of the total apple flowvalue to Nicosia.

The highest average annual precipitation in the entire Island falls on these leading apple flowvalue villages (Figure 29). Perennial mountain springs and streams provide abundant high quality low chloride water for gravity irrigation of apple orchards. Deciduous fruit trees grow well on terraced valley slopes on the alluvial and silicate raw soils which overlie the igneous rocks of the Troodes Massif.

Kilani and Platres, with their advantages of southern slope exposure, are able to harvest their apples early in the season during September and October (Table 13), and consequently obtain higher prices in the Nicosia Wholesale Market during November and December (Figure 31). Cold store refrigeration facilities in Nicosia allow merchant-dealers and cooperatives to release supplies gradually over the high demand, high price off-season running through the months of March, April, and May. An estimated 9 percent of the total apple flowvalues to the Nicosia Wholesale Market originated from cold storage facilities located within the Nicosia municipality.

High flowvalues for apples grown in the high altitude supply

---

[1]The aquifers of the Kokkinochoria peninsula, like those of the Florida peninsula, are recharged by rainfall, and are subject to overdraft and seawater infiltration. The problem is particularly acute during periods of drought, as this hydro-climatic tomato zone lies within the west-east trough of lowest average annual precipitation (Figure 26). See Toufexis and Jacovides (1972) for a complete hydrologic assessment of Cyprus for the 1969-70 period.

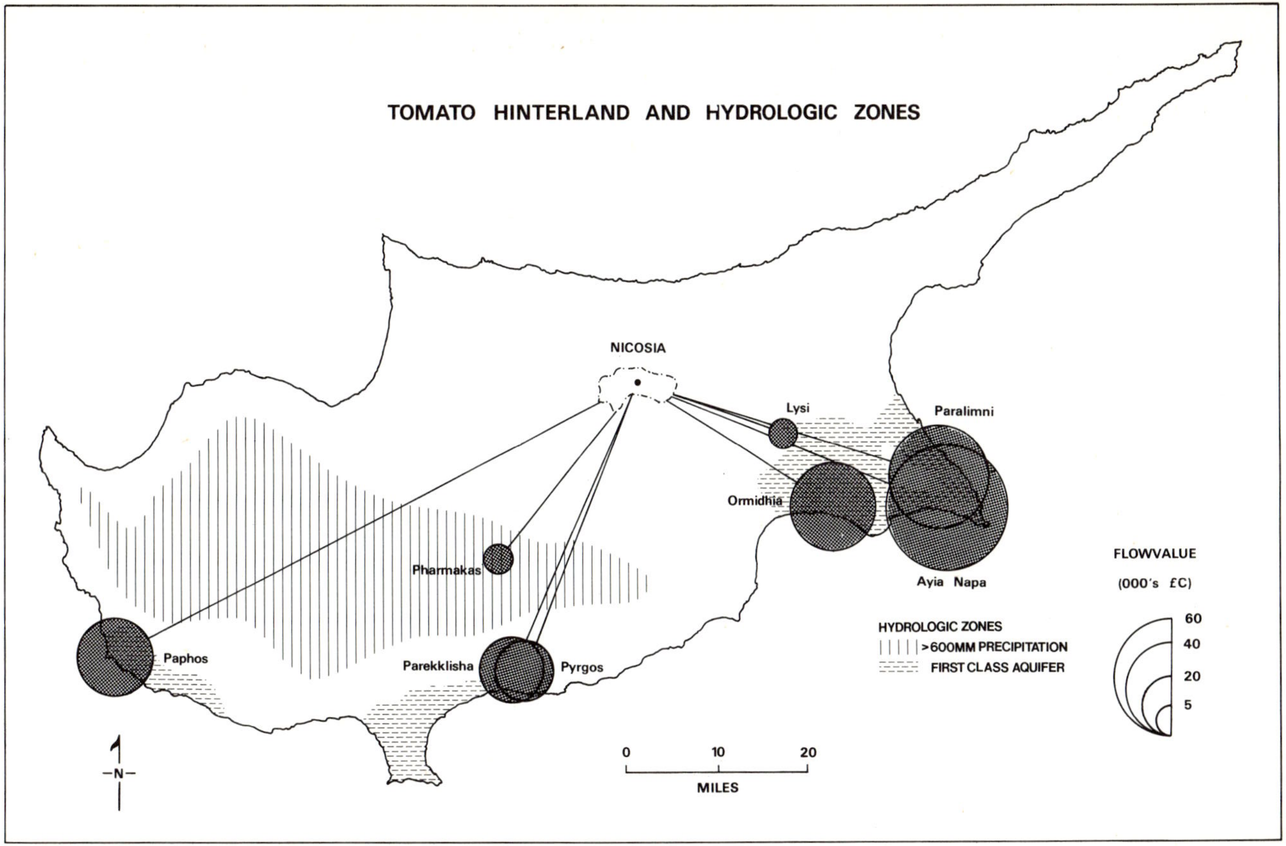

Fig. 28.--Tomato Hinterland and Hydrologic Zones

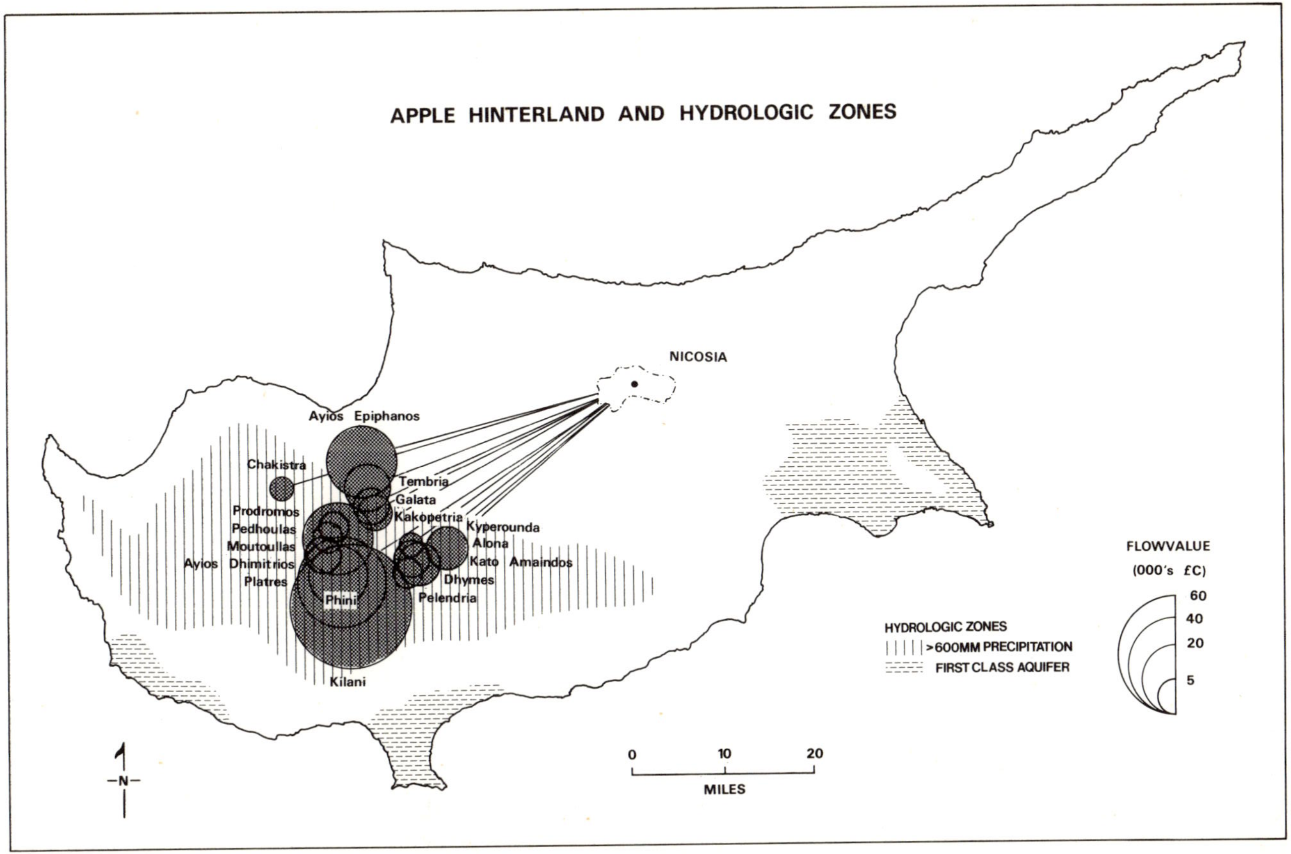

Fig. 29.--Apple Hinterland and Hydrologic Zones

TABLE 12

MONTHLY LABOR REQUIREMENTS FOR TOMATOES GROWN UNDER COVER (hrs/donum)

Operation	Month											
	J	F	M	A	M	J	J	A	S	O	N	D
Land preparation	–	–	–	–	–	–	–	0.5	–	0.5	0.5	–
Planting	–	–	–	–	–	–	–	–	–	–	–	8.0
Tillage	7.0	21.0	–	–	–	–	–	–	–	–	–	8.0
Fertilization	–	1.0	–	–	–	–	–	–	–	–	–	1.0
Irrigation	8.0	12.0	12.0	4.0	–	–	–	–	–	–	2.0	6.0
Plant protection	2.0	3.0	3.0	1.0	–	–	–	–	–	–	–	1.0
Constr. of low tunnels	–	–	–	–	–	–	–	–	–	–	–	72.0
Covering and Uncovering	10.0	20.0	20.0	–	–	–	–	–	–	–	–	–
Harvesting	–	–	80.0	80.0	40.0	–	–	–	–	–	–	–
Miscellaneous	2.3	4.6	9.2	6.8	3.2	–	–	–	–	–	0.2	7.7
Total	29.3	61.6	124.2	91.8	43.2	–	–	0.5	–	0.5	2.7	103.7

Source: Papachristodoulou (1970).

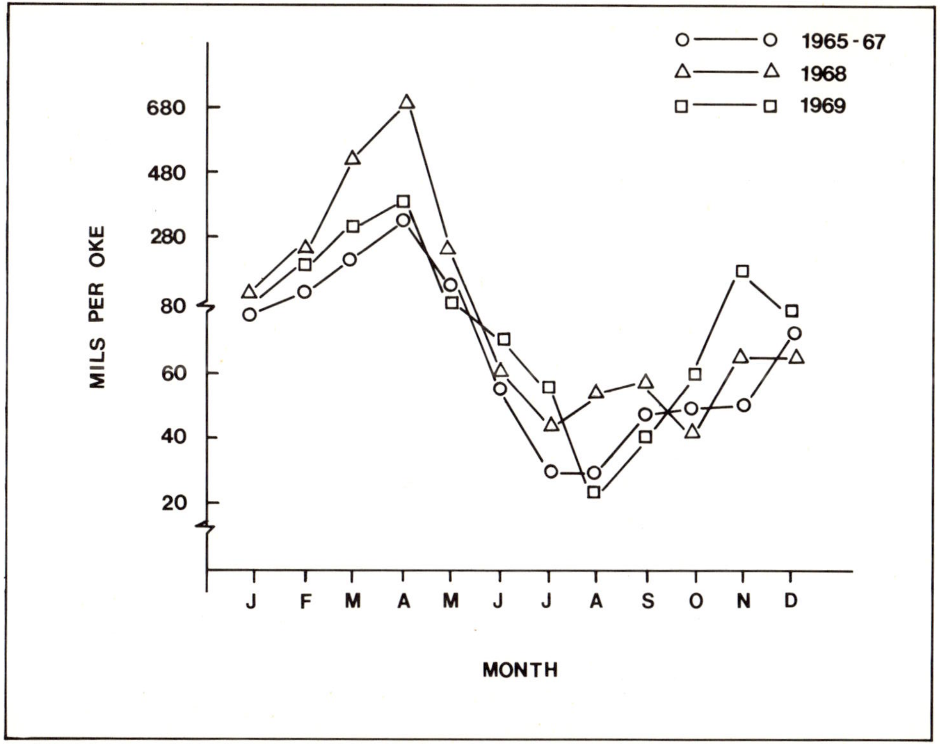

Fig. 30.--Average Monthly Prices of Tomatoes Sold in the Nicosia Municipal Wholesale Market 1965-69. Adapted from Papachristodoulou (1970).

TABLE 13

MONTHLY LABOR REQUIREMENTS FOR APPLES (hrs/donum)

Operation	Month											
	J	F	M	A	M	J	J	A	S	O	N	D
Tillage	–	3.0	–	15.0	–	–	15.0	–	–	–	3.0	–
Fertilization and manuring	7.5	–	–	2.0	2.0	–	–	–	–	–	2.0	–
Irrigation	–	–	–	–	2.0	2.0	2.0	4.0	2.0	–	–	–
Plant protection	4.0	–	4.0	4.0	8.0	8.0	4.0	4.0	–	–	4.0	–
Pruning	–	25.0	–	–	–	–	–	–	–	–	–	–
Harvesting	–	–	–	–	–	–	–	–	80.0	40.0	–	–
Miscellaneous	1.5	3.5	0.5	2.5	1.5	1.0	2.5	1.0	9.7	4.8	1.5	–
Total	13.0	31.5	4.5	23.5	13.5	11.0	23.5	9.0	91.7	44.8	10.5	–

Source: Papachristodoulou (1970).

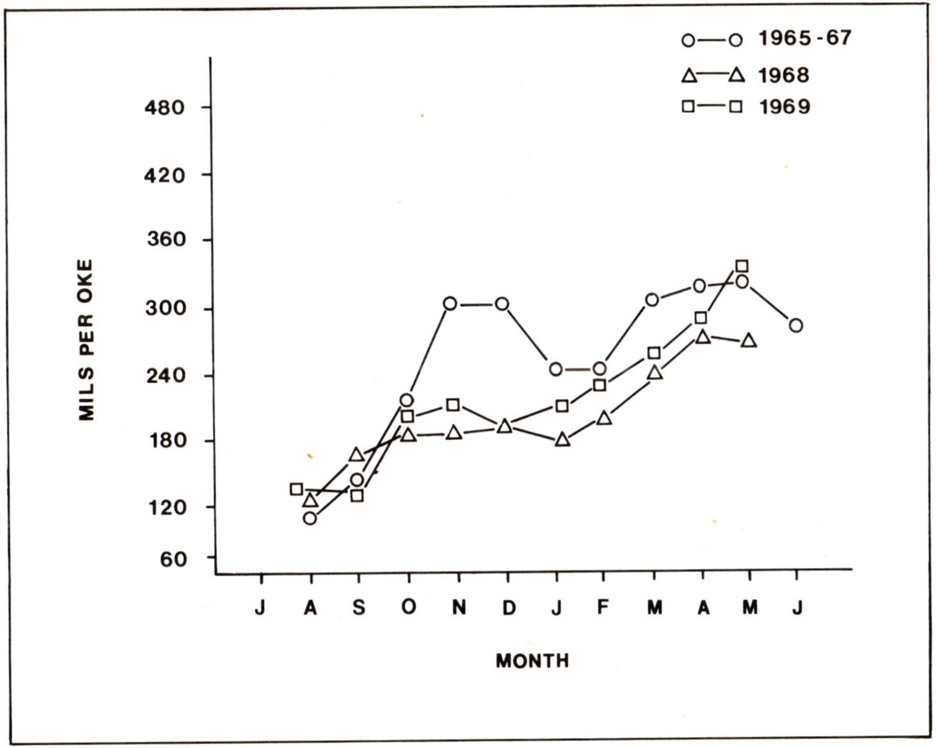

Fig. 31.--Average Monthly Prices of Apples Sold in the Nicosia Municipal Wholesale Market 1965-69. Adapted from Papachristodoulou (1970).

villages of the Troodos Mountains constitute a marked contrast to most of the mountainous areas of Latin America and Africa which generally are not important hinterlands for staple produce sent to urban markets (V. Smith 1975).

The hydrologic dichotomy of the staple produce supply regions of Cyprus should not obscure the basic underlying dependency of the irrigated lowlands upon the rainfed highlands. The orographic winter precipitation generated when traveling low pressure systems are forced to rise over the Troodos Mountains is the primary source of fresh water recharge for the mountain springs, spate reservoirs, and groundwater aquifers which irrigate the piedmont coastal plains skirting the Troodos Massif, and the Kokkinochoria truck farms.

Thus, drought not only affects the production of apples in the rainfed highlands, but also has an adverse effect on the replenishment of underground water resources for tomato cultivation in the lowlands.

Average annual precipitation of 495 mm provides almost all of the island's water resources. It ranges from in excess of 1,200mm in the higher elevations of the Troodos Mountains to less than 200mm in parts of the Mesaoria trough.

> In an average year, an estimated one million-million gallons of water falls on the island to re-supply rivers and groundwater beds. But, of this amount, only about 73 million gallons becomes usable surface runoff; more than half of the runoff--40,000 million gallons--flows into the sea lost to water-short villages (<u>Cyrpus Today</u> 1967, p. 10).

As Christoudoulou (1959, p. 21) points out, the seasonal distribution of precipitation in Cyprus' semi-arid eastern Mediterranean climate is of profound significance to horticultural production:

> The winter rainfall is probably the most important single element in the physical environment of Cyprus. The pattern of life is keyed to it. The life-cycle of vegetation is decided by it and man has had perforce to adapt farming practices accordingly.

Fresh water for irrigation is the lifeblood of staple food production in Cyprus, and is the key variable in determining the spatial structure of horticultural land use and commodity flows.

<u>Ricardo vs Thünen</u>

We conclude this analysis of the spatial structure of the produce supply hinterland with a statistical test of the Ricardo and Thünen approaches to horticultural land use. To reiterate, our initial hypothesis states that the regional specialization of major produce supply is a simultaneous function of hydrologic, climatic, and market distance variables. In the following section, the operational definitions of these three variables are given relative to tomato and apple flowvalue estimation.

Variable Selection

Hydrology. Comparative advantage in hydrologic resources for tomato supply areas is measured by donums of area irrigated through groundwater extraction. Irrigation data for fifty-two tomato supply origins were obtained from Toufexis and Jacovides (1972).[1] A positive relationship is hypothesized between tomato flowvalue and groundwater irrigation.

Hydrologic advantage for apple regionalization is expressed in average annual millimeters of precipitation for 90 supply villages Data were obtained from Cyprus (1972b). As the level of average annual precipitation increases, the value of apple commodity flow in expected to increase concomitantly.

Climate. Reliable temperature and humidity data on a scale suitable for this study were unavailable. Consequently a proxy measure of climatic advantage was substituted to reflect the particular requirements of tomato and apple cultivation. Although the bias of the estimates may tend to be larger with coastal distance than with the true climatic variables, the extent of bias is assumed to be negligible for test purposes (Rao and Miller 1971, pp. 82-88).

It has been noted that the moderating influence of the sea on low temperature extremes during winter results in an earlier and longer frost-free growing season, and ultimately, in higher prices for early season tomatoes in the central wholesale market. Moreover, warmer annual and diurnal nyctotemperature together with higher relative humidity constitute further natural advantages to locating tomato truck farms near the warm coastal waters of southern Cyprus.

In view of the long term temperature and humidity benefits accruing from coastal proximity, it was decided to operationally define climatic advantage in terms of the shortest straight-line distance in statute miles from the administrative boundary of the tomato supply origin to the nearest seacoast (Data source: Cyprus 1973b).

---

[1]Data availability dictated the use of a sample of 52 supply origins for the test of tomato flowvalue determinants (Table 5). Whereas, the original sample of 90 supply villages is used in testing the apple supply hypothesis.

Tomato flowvalue is expected to be inversely proportional to coastal distance.

In contrast to the tomato's hypothesized proclivity to coastal proximity, the apple prefers cooler temperatures associated with higher elevation and consequent greater distance inland as well as upland from the coast. Therefore, it is hypothesized that apple flowvalue is directly related to distance from the village boundary to the seacoast.

Market Distance. Market distance is defined as the shortest route in statute miles from the produce supply origin to the Nicosia market via the bituminous road network (Data source: Cyprus 1973b).

Despite the low degree of positive direction of association evinced between market distance and aggregate flowvalue, it was decided to retain the distance to market variable along with the hydrologic and climatic predictors in the final regression models. Thereby, enabling us to formally "play-off" the Thünen (market distance) theory of land use against the Ricardo (land productivity) explanation. On one hand, the Thünen view suggests that the closer a supply village is to the market, the more intense the cultivation of the land, and the higher the value of produce flow from that village to market. On the other hand, the Ricardo approach proposes that the more environmentally productive the village land, the more intensive the use, and the higher the flowvalue.

These hydrologic, climatic, and market distance variables are regressed on tomato and apple flow scores to establish the relative importance of environmental factors and market distance in predicting the observed pattern of regional specialization.

Regression Model

The multiple regression technique[1] with forward stepwise inclusion is employed to search out whether the Thünen or the Ricardo view of economic rent makes the greatest contribution toward explaining the total variation in tomato and apple flowvalues.

The general multiple regression model is in the form

$$Y = a + b_1X_1 + b_2X_2 + b_3X_3 + e$$

---

[1]SPSS subprogram REGRESSION with forward (stepwise) inclusion is used in both tomato and apple analyses (Nie et al. 1975, pp. 320-367). Yeates (1974, pp. 120-122) summarizes the conditions under which the use of the stepwise multiple regression technique is appropriate.

where Y is the dependent variable, a is the regression hyperplane intercept with Y, the b's are the regression coefficients, the X's are the explanatory variables, and e is the error term. This general model is then adapted to the specific research hypotheses for tomato and apple flowvalue determinants.

Collectively, the first operational hypothesis states that tomato flowvalue is directly related to the volume of groundwater irrigation, and distance to market; and inversely related to distance to the seacoast.

The raw data for each variable are positively skewed and have been logarithmically transformed to approximate normality and homoscedasticity.[1] The form of the multiple least-squares equation now becomes

$$\log_{10}T = \log_{10}a + b\log_{10}G - b\log_{10}C + b\log_{10}M$$

where T is tomato flowvalue, a is the intercept, G is groundwater irrigation, C is coastal distance, and M is market distance. Substituting T' for $\log_{10}T$, a' for $\log_{10}a$, G' for $\log_{10}G$, and M' for $\log_{10}M$, we obtain the prediction equation

$$T' = a' + bG' - bC' + bM'$$

Table 14 summarizes the overall results of the stepwise regression estimation of tomato flowvalue. The calculated equation (with standard error of the estimated b) at step 3 is

$$T' = 1.6267 + \underset{(0.1918)}{0.7104G'} - \underset{(0.2941)}{0.6388C'} + \underset{(0.8282)}{0.4633M'}$$

$$R^2 = 0.337$$

The positive regression coefficient for G', and the negative coefficient for C' are both significant at the 0.05 level of confidence on the basis of an F-test with 48 degrees of freedom. The regression coefficient for M' is smaller than the standard error, and is not significant at the 0.05 level. Thus, the first research hypothesis is not as a whole substantiated by the simultaneous test of the three parameters at step 3.

However, at step 2, the two environmental variables, area irrigated (G') and coastal distance (C'), combine to account for one-third of the variance in tomato flowvalue (T'). Both regression coefficients are significant at the 0.05 level (d.f. = 49),

---

[1]The transformed data for the tomato and the apple flowvalue analyses are listed in Appendices B and C, Tables 17 and 18 respectively.

TABLE 14

STEPWISE REGRESSION ESTIMATION OF TOMATO FLOWVALUE

Zero-Order Correlation Matrix

	Tomato	Area Irrigated	Coastal Distance	Market Distance
Tomato	1.000	0.508	-0.384	0.408
Area Irrigated		1.000	-0.260	0.352
Coastal Distance			1.000	-0.798
Market Distance				1.000

Multiple Correlation Coefficients

Step Number	Variable Entered	R	R^2	R^2 Change
1	Area Irrigated	0.508	0.258	0.258
2	Coastal Distance	0.571	0.326	0.068
3	Market Distance	0.581	0.337	0.011

Regression Statistics at Step 3

Explanatory Variable	Regression Coefficient	Standardized Regression Coefficient	Standard Error	F Value
Area Irrigated	0.7104	0.4604	0.1918	13.721
Coastal Distance	-0.6388	-0.2647	0.2941	4.719
Market Distance	(0.4633)	(0.1127)	0.8282	(0.310)

Estimation Summary at Step 3

R	0.581
R^2	0.337
Adjusted R^2.	0.296
Standard Error	1.092
F Value.	8.136
Sample Size.	52

Coefficient and F value in parentheses are not significant at the .05 level of confidence.

and the multiple coefficient of determination is significant at the 0.01 level. The result at step 2, where market distance is not entered in the equation, demonstrates that tomato flowvalue does indeed vary directly with the area irrigated by groundwater extraction, and inversely with distance to the seacoast. Figure 32 shows the direct relationship between tomato flowvalue and groundwater irrigation.

The second operational hypothesis stipulates that apple flowvalue is directly related to average annual precipitation, distance to the seacoast, and distance to market.

Raw data were transgenerated to common logarithms, and the parameters of the following equation were estimated:

$$\log_{10}A = \log_{10}a + b \log_{10}P + b \log_{10}C + b \log_{10}M$$

where A is apple flowvalue, P is precipitation, C is coastal distance, and M is market distance. Substituting A' for $\log_{10}A$, P' for $\log_{10}P$, C' for $\log_{10}C$, and M' for $\log_{10}M$, the prediction equation becomes

$$A' = a + bP' + bC' + bM'$$

The results of the estimate are given in Table 15. The calculated equation at step 3 is

$$A' = -10.4443 + \underset{(0.7410)}{3.3819P'} + \underset{(0.1928)}{0.8378C'} + \underset{(0.3279)}{1.2820M'}$$

$$R^2 = 0.633$$

The positive regression coefficients for P', C', and M' are all significant at the 0.001 level (d.f. = 86), and the R^2 (also significant at the 0.001 level) indicates that 63 percent of the variation in apple flowvalue is explained by the combined influence of precipitation, coastal distance, and market distance. The overall relationship is direct, confirming the validity of the second operational hypothesis. Figure 33 illustrates the positive relationship between apple flowvalue and precipitation.

In both the tomato and apple flowvalue estimates, the largest residuals from the regression line are due to data inaccuracies and lack of standardization in those predictor variables derived from cartographic analysis. Scale and measurement error, chronic afflictions in obtaining data through map interpretation, are particularly troublesome in estimating the value of fresh produce production because the regressand is so acutely sensitive to micro-scale variations in hydrology and climate. Consequently, interpretation emphasis is placed on the direction of relationship and on the relative importance of individual regressors, rather than on the overall

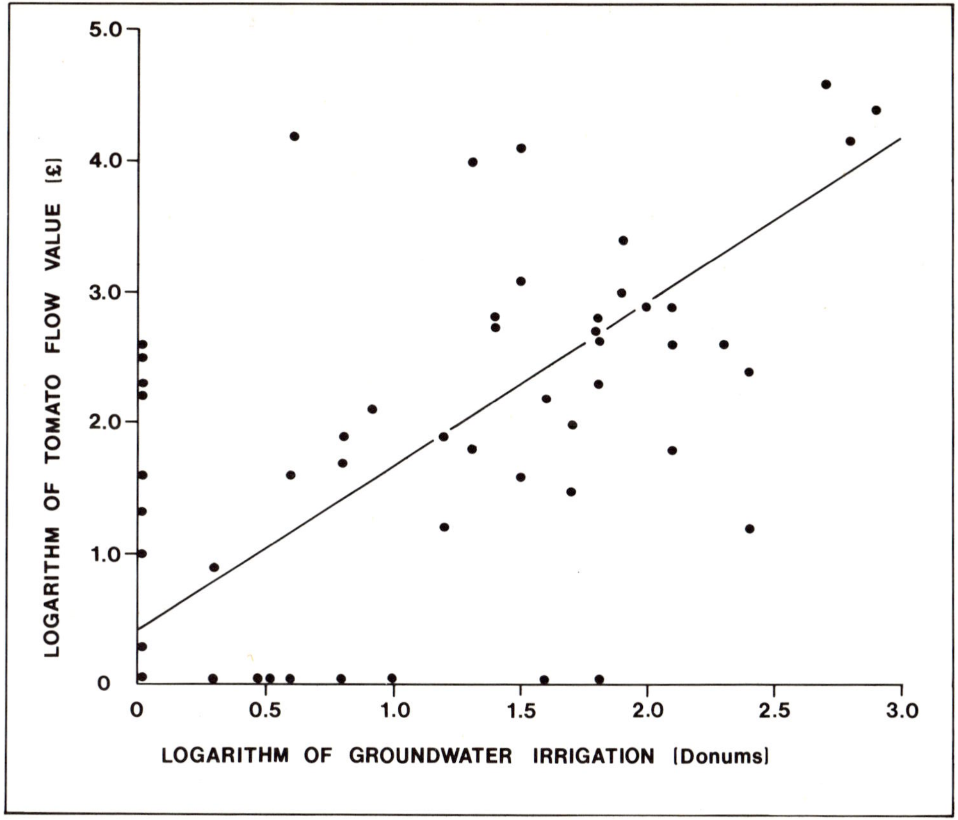

Fig. 32.--Relationship between Tomato Flowvalue and Groundwater Irrigation.

TABLE 15

STEPWISE REGRESSION ESTIMATION OF APPLE FLOWVALUE

Zero Order Correlation Matrix

	Apple	Precipitation	Coastal Distance	Market Distance
Apple	1.000	0.731	0.415	0.505
Precipitation		1.000	0.335	0.540
Coastal Distance			1.000	-0.216
Market Distance				1.000

Multiple Correlation Coefficients

Step Number	Variable Entered	R	R^2	R^2 Change
1	Precipitation	0.731	0.535	0.535
2	Coastal Distance	0.753	0.567	0.032
3	Market Distance	0.795	0.633	0.065

Regression Statistics at Step 3

Explanatory Value	Regression Coefficient	Standardized Regression Coefficient	Standard Error	F Value
Precipitation	3.3819	0.4247	0.7410	20.828
Coastal Distance	0.8378	0.3484	0.1928	18.883
Market Distance	1.2820	0.3510	0.3279	15.286

Estimation Summary

R	0.795
R^2.	0.633
Adjusted R^2	0.620
Standard Error. . .	0.798
F Value	49.350
Sample Size . . .	90

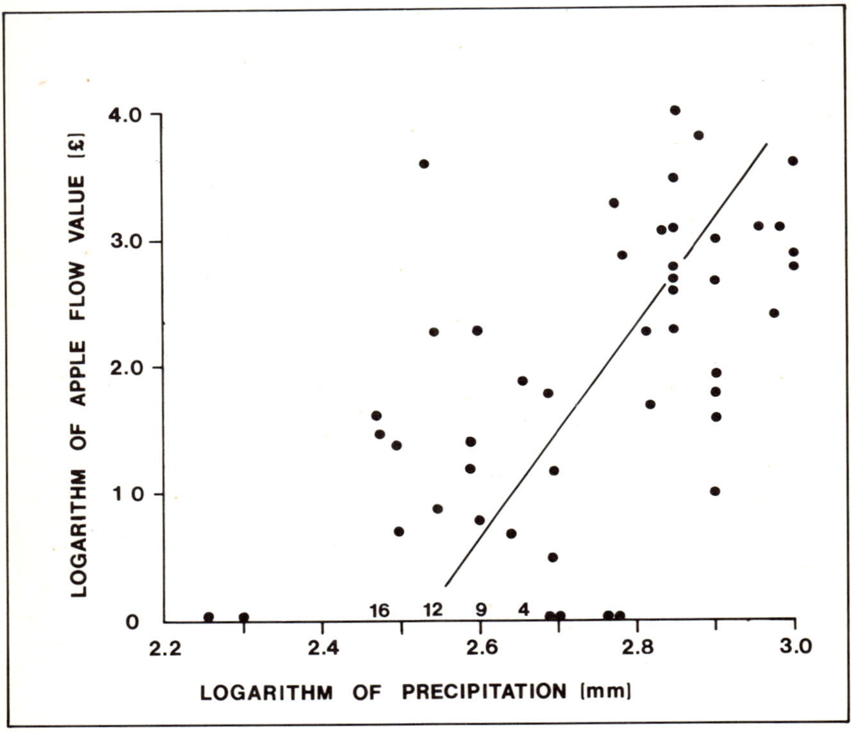

Fig. 33.--Relationship between Apple Flowvalue and Precipitation.

predictive power of the equations.

Assessment

We now turn to an evaluation of the relative importance of the hydrologic, climatic, and market distance variables by examining the standardized regression coefficients of the tomato and apple equations at step 3.[1]

$$T'_s = .4604G' - .2647C' + .1127M'$$

$$A'_s = .4247P' + .3484C' + .3510M'$$

The standardized regression coefficients, or beta weights, show how much change in Tomato and in Apple flowvalue is produced by a standardized unit change in one of the three independent variables when the other two are held constant (Blalock 1972, p. 453).

In the tomato flowvalue equation, the result indicates that the relative importance of groundwater irrigation far outweighs both coastal distance and market distance. And, the beta weights in the apple flowvalue equation show that apple commodity flows to the central market are primarily related to the level of precipitation which keeps the mountain springs and streams flowing for irrigation. Again, as with tomato flows, coastal distance and market distance are of secondary importance.

In brief, the overall results of the two regression analyses support the Ricardo approach to horticultural location through environmental factors, and point to an underlying principle of hydrologic primacy which shapes the pattern of regional specialization in the prodcue supply hinterlands of Cyprus.

[1]The regression coefficients are standardized by multiplying each coefficient by the ratio between the standard deviation of the variable to which the regression coefficient pertains and the standard deviation of the dependent variable (Yeates 1974, p. 121).

CHAPTER IV

CONCLUSION

Summary of Findings

Staple produce commodity flows from a sample of ninety supply origins to thirty-eight wholesaler destinations in the Nicosia Municipal Wholesale Market were examined to test two general research hypotheses.

The 90 x 38 flowvalue matrix was subjected to principal components analysis to test the first proposition that shared patterns of flow linkage between groups of wholesaler destinations in the central market and common sets of supply origins in the hinterland can be identified. Results indicate that nearly three-fifths of the total variance in the 38 x 38 correlation matrix was accounted for by four distinct groups of wholesalers.

By matching wholesaler group factor loadings with their respective supply origin factor scores, it was found that a group of twelve merchant dealers (24% of total variance) obtain primarily tomatoes and cucumbers from a set of three main supply origins including the Paphos area, and the villages of Ormidhia and Pygros. A second group (14% of total variance), consisting of one cooperative and six merchant dealers wholesales tomatoes and cucumbers grown in Ayia Napa and Ormidhia. The third group of six independent truckers (11% of total variance) share inflows of watermelon from Zohhia and Morphou. And, the fourth group of three independent truckers (7% of total variance) wholesales apples and watermelon originating in Nicosia's cold storage facilities.

An equally important finding was that the remaining two-fifths of the total variance is attributable to bipolar and unique flowvalue linkages. This means that a single source of supply is shared with only one other wholesaler (bipolar), or it is the exclusive domain of only one wholesaler (unique).

An outstanding example of this form of dyadic specialization in which one wholesaler dominates the disposal of produce from one major supplier is the Paralimni situation. This village is the third highest value supplier of fresh food to Nicosia. And, over ninety percent of its staple produce flowvalue is wholesaled through one agent--the Paralimni Growers' Cooperative Organization.

Larger cooperatives such as Paralimni and Ayia Napa tend to monopolize both the supply and the wholesale marketing of staple produce grown in their respective village areas.

Our second research hypothesis addressed the spatial structure of produce supply in the hinterland. The working proposition was that regional specialization of staple produce supply origins is a simultaneous function of environmental resources and market distance. This hypothesis was tested by stepwise multiple regression analyses of two data sets--one for each of the two highest value commodities brought to the central wholesale market (tomatoes and apples). The objective of the analyses was to determine the relative importance of the Ricardo and Thünen approaches to the explanation of horticultural specialization in the central market hinterland.

Employing a 52 x 4 data matrix, groundwater extraction for irrigation, distance to the seacoast, and distance to market were regressed on tomato flowvale scores for fifty-two sample villages. The regression estimate indicates that tomato flowvalue is directly related to the area irrigated by groundwater extraction, and inversely related to distance to the seacoast. Market distance was not a significant predictor at the 0.05 level of confidence.

Using a 90 x 4 data matrix, average annual precipitation, distance to the seacoast, and distance to the central wholesale market were regressed on apple flowvalue scores for ninety sample villages. Regression results confirm that apple flowvalue is directly related to average annual precipitation, coastal distance, and market distance.

In both the tomato and the apple regression analyses, the relative importance of the hydrologic environmental predictor is outstanding. Thus, we conclude that the Ricardo view of economic rent makes the greatest contribution to explaining the pattern of regional specialization in tomato and apple hinterlands.

With regard to sources of supply in general, one important limitation of this study is that we have dealt with the value of produce flows summed over the course of an entire year. Our use of aggregate annaual flowvalues tends to emphasize the distribution of higher-value out-of-season producing areas. We may thereby be masking marked seasonal migrations in the geography of horticultural production in Cyprus.

In view of this constraint, it appears that a topic ripe for further study would be the extent and cause of seasonal fluctuations in horticultural supply areas. Citing the case of cucumber supply in particular, Market Inspector Ferekides suggests that there is considerable seasonal movement of produce supply areas in Cyprus.[1]

Holding water availability spatially constant (a very critical presumption), we may speculate that during cooler, out-of-season periods of produce scarcity and high price, major supply areas are found furthest from market near the frost-free southern coastlands. Whereas, during warmer normal season periods of abundance and low price, production becomes possible closer to market and thereby its location becomes more sensitive to transport cost.

For certain crops, such as cucumbers, inward contraction of supply areas toward the market may occur during warmer weather. This may then alternate with outward expansion away from the market during cooler weather. In effect, this conjectural pattern reduces to a pulsating seasonal rhythm similar to that found in the local supply of fresh produce for California (Gregor 1957), and the national supply of fresh vegetables for the United States (Jumper 1969).

A second limitation of this study arises from our focusing on the central wholesale market to take the pulse-beat of the entire internal wholesale produce marketing system. A questionnaire survey of the four smaller municipal wholesale markets (Famagusta, Limassol, Larnaca, Paphos) and one village wholesale market at Morphou, was well underway when the Turkish invasion stopped us short in our field research. Completion of this survey would have provided us with the necessary data with which to evaluate similarities and differences in marketing practices among all six wholesale produce markets in the system.

Synthesis

Perhaps the most important finding of this study is the dominant theme of increasing value of fresh produce flow with increasing distance from the central wholesale market.

This direct value-distance relationship does not agree with the conclusions drawn in similar studies of produce supply areas by Thünen (1826), Eyre (1959), Hunt (1959), and D. Smith (1966). It is, however, in accordance with the body of agricultural location theory developed by Ricardo (1817), Baker (1921), Brinkman (1935), Chisholm (1962), and Jumper (1969).

Apparently, market distance has taken a back seat to environmental variables in the truck-farming industry. The extension of

---

[1]Personal communication with S. Ferekides, Inspector of Wholesale Markets, summer, 1974.

bituminous roads to distant irrigated areas of Cyprus, together with the shift in transport mode from donkey-back to heavy lorry, has reduced the friction of distance to the extent that the annual value of produce supply increases outward from the central wholesale market. In short, regardless of market distance, high value major produce supply areas in Cyprus are now concentrated in those remote parts of the hinterland which offer the most favorable hydrologic conditions. Availablity of fresh water for irrigation was found to be the key variable which shapes the pattern of regional specialization in fresh food supply for the Nicosia Municipal Wholesale Market.

On a broader level, a marked cross-cultural similarity exists between Cyprus and the United States in the geography of fresh food supply. Despite the obvious contrasts in scale and economic base, there is a fundamental parallel in the spatial pattern of high value horticultural supply for major urban markets, and in the process behind the pattern.

In both countries, the highest value fresh vegetable supply regions are located along the southern periphery of the land mass, on irrigated lowlands, and in close proximity to the sea. Early season crops grown under cover on these frost-free truck farms command high prices in northern urban markets where demand is highest.

The parallel may be pushed a bit further. In both the United States and Cyprus, the focus of high demand for fresh produce in larger urban centers served by efficient overland transport has stregthened the position of urban wholesalers in the marketing chain.

It is clear that merchant dealers and cooperative organizations are the prime movers in Nicosia's wholesale produce marketing chain. Through their commission services to local dealers, cooperative agents, and growers, merchant dealers ultimately handle ninety-seven percent of the total value of staple produce flows to the central wholesale market. Moreover, they set wholesale prices for the entire internal marketing system of Cyprus.

The urban wholesalers' dominance of the internal marketing system is augmented by the high level of consumer demand for fresh food in greater Nicosia, and by Nicosia's eminent accessibility as the hub of the Island's bituminous road net. The ease of relatively long distance bulk transport by truck enables Nicosia's merchant dealers to tap all productive parts of the hinterland. And, small-scale produce farmers are thereby given incentive for expanded production for market. In addition, growers' cooperatives make it possible for many small farmers to take advantage of economies of scale in the collection, shipment, and sale of their fresh food crops.

In terms of a general body of theory on the evolution of internal marketing systems, the results of this study essentially corroborate the findings of Hodder and Ukwu in West Africa (1969), Forman and Rieglehaupt in Northeast Brazil (1970), William Jones in Tropical Africa (1972; 1974), Norton and Symanski in Jamaica (1975) and V. Smith in Ecuador (1975). A common theme running through each of these previous works is succinctly summarized by Forman and Rieglehaupt:

> ...the marketing system will lead to a restructuring of the production system when the latter is unable to meet consumer demand.

Elaboration on this theme through our analysis suggests that increased urban demand for fresh produce combined with improvements in transportation and market accessibility have had a twofold impact on the internal produce marketing system of Cyprus. First, it enhanced the dominant role of urban wholesalers in the marketing chain. And, second, it facilitated a spatial shift outward from market-proximate gardening toward specialized truck-farming in remote areas of the hinterland which have greatest comparative advantages in hydrologic resources.

APPENDIX A

MAJOR PRODUCE SUPPLY ORIGINS WITHIN THE TURKISH OCCUPIED ZONE

TABLE 16

MAJOR PRODUCE SUPPLY ORIGINS WITHIN THE TURKISH OCCUPIED ZONE

Supply Origin	Village or Town	Census Code Number	Flowvalue 1970 (Pounds Cyprus)	% of Total Flowvalue Behind the Attilla Line	Cumulative %
47	Kondea	343	£ 11378	30.31	30.31
49	Lysi	354	6508	17.34	47.65
35	Zodhia	178	3611	9.62	57.27
40	Akanthou	302	2896	7.71	64.98
9	Ayia Marina	22	2143	5.71	70.69
42	Athna	311	2004	5.34	76.03
22	Morphou	114	1796	4.78	80.81
41	Asha	310	1745	4.65	85.46
39	Famagusta	301	1271	3.39	88.85
37	Kalogrea	215	1072	2.86	91.71
27	Pendayia	132	529	1.41	93.12
38	Lapithos	230	481	1.28	94.40
18	Kato Kopia	75	391	1.04	95.44
14	Galini	54	360	.96	96.40
8	Aviana	19	298	.79	97.19
5	Argaka	15	266	.71	97.90
46	Kalopsidha	338	264	.70	98.60
30	Philia	140	146	.39	98.99
36	Kyrenia	201	124	.33	99.32
44	Ayios Evstathio	318	108	.29	99.61
52	Sinda	381	95	.25	99.86
54	Vatili	393	53	.14	100.00
	Total 22 Supply Villages		£ 37539		

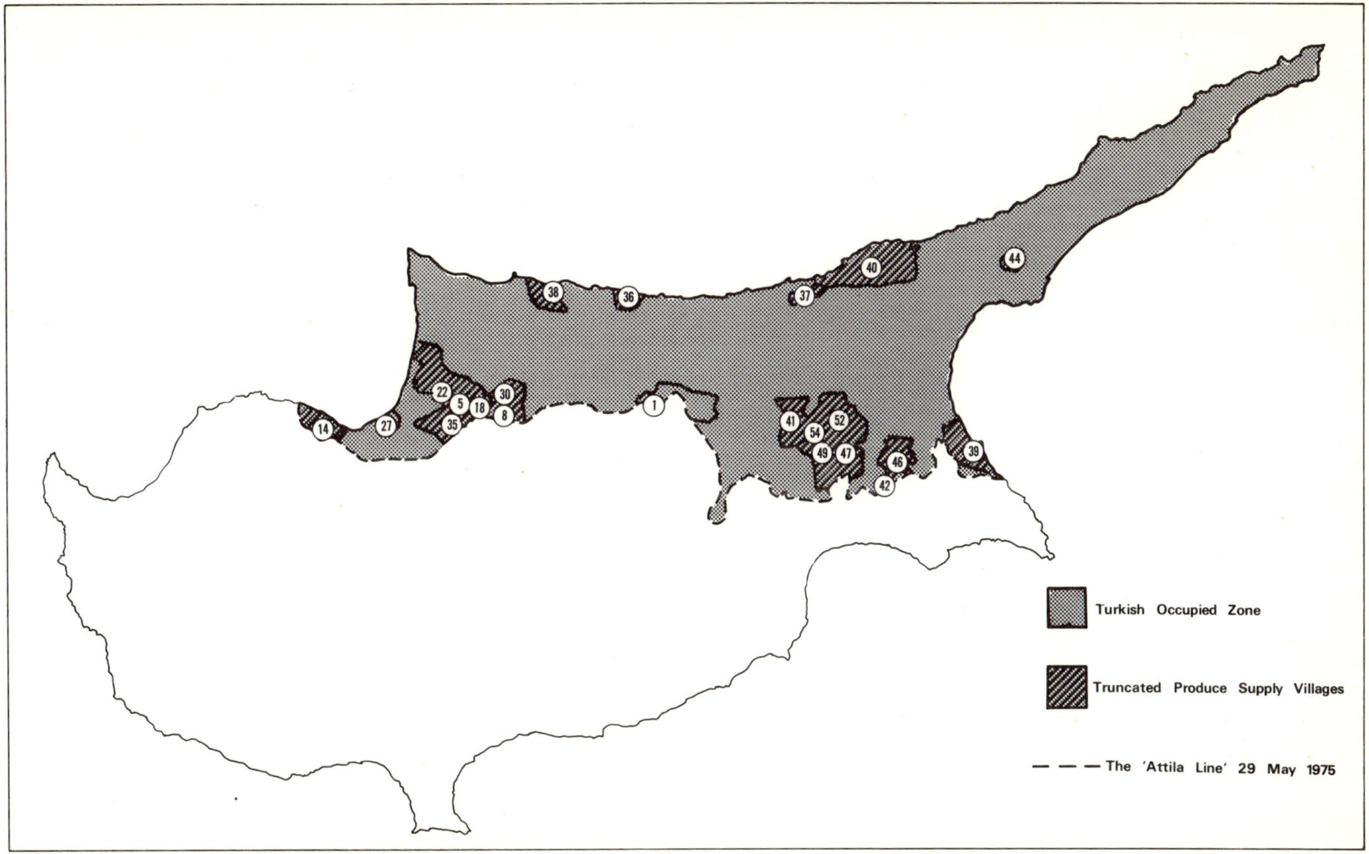

Fig. 34.--Portion of the Nicosia Produce Hinterland Truncated by the Turkish Invasion

APPENDIX B

SUPPLY ORIGIN DATA FOR TOMATO FLOWVALUE ANALYSIS

TABLE 17

SUPPLY ORIGIN DATA FOR TOMATO FLOWVALUE ANALYSIS
(All Data Logarithmically Transformed)

Origin Number	Tomato Flowvalue (£000's)	Market Distance (Miles)	Coastal Distance (Miles)	Area Irrigated (Donums)
1	2.477	1.176	1.230	0.0
2	2.959	1.041	1.114	2.173
3	0.0	1.322	0.845	1.477
4	0.0	0.954	1.301	0.778
5	1.000	1.230	0.903	0.0
6	0.0	1.230	1.041	1.079
7	3.080	1.806	0.0	1.944
8	0.0	1.903	1.255	1.839
9	2.556	1.613	0.0	0.0
10	1.342	1.279	0.903	0.0
11	2.746	1.000	1.176	1.820
12	0.0	0.602	1.204	0.301
13	2.769	1.000	1.114	1.398
14	2.336	1.398	0.0	0.0
15	2.827	1.079	1.204	1.881
16	2.137	1.519	0.0	0.954
17	2.767	1.041	1.342	1.447
18	0.0	1.204	1.000	0.477
19	1.875	1.041	1.362	0.845
20	1.633	0.477	1.114	0.0
21	0.0	1.301	0.778	0.477
22	0.301	1.204	0.0	0.0
23	2.255	1.519	0.0	0.0
24	0.954	1.663	0.0	0.301
25	1.959	1.580	0.0	0.778
26	1.908	1.580	0.0	1.230
27	1.699	1.146	1.146	0.845
28	1.643	1.505	0.0	1.556
29	4.646	1.708	0.0	2.673
30	2.562	1.623	0.0	2.072
31	0.0	1.415	1.041	0.0
32	3.096	1.342	0.954	1.462
33	1.176	1.699	0.477	1.204
34	3.418	1.301	0.0	1.934
35	4.471	1.663	0.0	2.825
36	1.556	1.663	0.778	1.708
37	2.662	1.663	0.0	1.857
38	0.0	1.415	0.0	0.602
39	2.954	1.477	0.0	1.987
40	2.265	1.519	0.602	1.857
41	1.613	1.531	0.301	0.602
42	1.987	1.544	0.0	1.732
43	2.167	1.519	0.0	1.643
44	4.358	1.591	0.0	2.782
45	1.771	1.556	0.0	1.301
46	1.230	1.633	0.0	2.386
47	4.109	1.708	0.0	1.477
48	4.007	1.699	0.0	1.301
49	4.231	1.973	0.0	0.602
50	2.601	1.826	0.0	2.314
51	1.813	1.851	0.0	2.158
52	2.391	1.813	0.0	2.356

APPENDIX C

SUPPLY ORIGIN DATA FOR APPLE FLOWVALUE ANALYSIS

TABLE 18

SUPPLY ORIGIN DATA FOR APPLE FLOWVALUE ANALYSIS
(All Data Logarithmically Transformed)

Origin Number	Apple Flowvalue (£000's)	Market Distance (Miles)	Coastal Distance (Miles)	Average Annual Precipitation (Millimeters)
1	0.0	0.0	1.041	2.498
2	0.0	1.176	1.230	2.477
3	0.0	1.041	1.114	2.477
4	3.099	1.477	1.279	2.845
5	0.0	1.322	0.845	2.477
6	0.0	0.954	1.301	2.544
7	0.0	1.230	0.903	2.477
8	0.0	1.230	1.041	2.477
9	0.0	1.806	0.0	2.653
10	3.581	1.602	0.778	2.544
11	2.660	1.681	1.000	2.903
12	0.0	0.903	1.255	2.544
13	2.984	1.491	1.041	2.778
14	0.0	1.613	0.0	2.602
15	3.069	1.505	1.079	2.813
16	2.458	1.690	1.000	2.954
17	1.643	1.663	1.000	2.903
18	0.0	1.279	0.903	2.477
19	1.544	1.000	1.176	2.477
20	0.0	0.602	1.204	2.544
21	0.0	1.000	1.114	2.301
22	0.0	1.398	0.0	2.477
23	2.901	1.699	1.041	3.000
24	0.0	1.740	0.0	2.653
25	0.0	1.079	1.204	2.477
26	2.769	1.716	1.079	3.000
27	0.0	1.519	0.0	2.462
28	0.0	1.041	1.342	2.544
29	1.000	1.380	1.230	2.916
30	0.0	1.204	1.000	2.477
31	2.292	1.602	0.845	2.544
32	0.0	1.041	1.362	2.602
33	0.0	0.477	1.114	2.544
34	3.270	1.568	1.000	2.778
35	1.623	1.301	0.778	2.477
36	0.0	1.204	0.0	2.740
37	0.477	1.519	0.0	2.699
38	1.204	1.663	0.0	2.699
39	1.176	1.580	0.0	2.591
40	0.0	1.580	0.0	2.699
41	0.0	1.146	1.146	2.498
42	0.0	1.505	0.0	2.498
43	1.398	1.708	0.0	2.591
44	0.0	1.690	0.699	2.602
45	0.0	1.623	0.0	2.591
46	0.0	1.415	1.041	2.544
47	0.699	1.342	0.954	2.498
48	0.0	1.699	0.477	2.531
49	0.0	1.301	0.0	2.498
50	0.0	1.663	0.0	2.565

TABLE 18--Continued

Origin Number	Apple Flowvalue (£ 000's)	Market Distance (Miles)	Coastal Distance (Miles)	Average Annual Precipitation (Millimeters)
51	0.0	1.663	0.778	2.571
52	0.0	1.322	1.114	2.544
53	0.0	1.663	0.0	2.591
54	0.0	1.415	0.0	2.544
55	0.0	1.415	0.0	2.544
56	0.0	1.477	0.0	2.544
57	0.845	1.519	0.602	2.602
58	0.903	1.531	0.301	2.544
59	0.0	1.531	0.845	2.602
60	0.0	1.544	0.0	2.544
61	0.0	1.519	0.0	2.544
62	1.740	1.415	0.0	2.813
63	1.431	1.591	0.0	2.498
64	0.0	1.556	0.0	2.544
65	0.0	1.279	1.079	2.778
66	0.0	1.633	0.0	2.498
67	0.699	1.732	0.0	2.653
68	2.599	1.875	1.255	2.845
69	3.008	1.934	1.255	2.903
70	3.067	1.968	1.255	2.954
71	0.0	1.301	1.114	2.778
72	3.142	2.000	1.279	2.845
73	2.354	1.978	1.255	2.845
74	4.048	1.919	1.079	2.845
75	2.701	1.964	1.204	2.845
76	1.954	1.748	1.146	2.903
77	0.0	1.708	0.0	2.653
78	2.757	1.973	1.204	2.845
79	3.480	1.944	1.204	2.845
80	3.815	1.934	1.176	2.875
81	3.618	1.732	1.255	3.000
82	1.903	1.699	0.0	2.653
83	2.332	1.892	1.079	2.813
84	1.806	1.973	1.176	2.903
85	2.265	1.973	0.0	2.602
86	0.0	1.826	0.0	2.602
87	1.778	2.025	0.778	2.699
88	0.0	1.851	0.0	2.602
89	0.0	1.771	0.0	2.602
90	0.0	1.813	0.0	2.653

SELECTED BIBLIOGRAPHY

Anderson, James R.
A geography of agriculture. The Brown Foundations of Geography Series. Dubuque, Iowa: Wm. C. Brown Company Publishers.

Azinas, Andreas
1973 Annual report, Department of Co-operative Development, years 1970, 1971, 1972. Nicosia: Department of Co-operative Development.

Baker, O. E.
1921 The increasing importance of the physical conditions in determining the utilization of land for agricultural and forest production in the United States. Annals, Association of American Geographers, 11:17-46.

Baldini, B.
1968 Cyprus. UN Food and Agriculture Organization of the United Nations, Commodity Bulletin 42. Horticulture in the Mediterranean Area: Outlook for Production and Trade. Rome: FAO.

Bear, L. M.
1963 Geological map of Cyprus. Geological Survey Department. Nicosia: Department of Lands and Surveys.

Belshaw, Cyril S.
1965 Traditional exchange and modern markets. Chapter 3: Monetized peasant marketing:53-58. Englewood Cliffs, N. J.: Prentice-Hall.

Berry, Brian J. L.
1966 Essays on commodity flows and the spatial structure of the Indian economy. Research Paper III. Chicago: The Department of Geography, The University of Chicago.

1967 Geography of market centers and retail distribution. Englewood Cliffs, N. J.: Prentice-Hall.

Berry, Brian J. L., Edgar C. Conkling and D. Michael Ray
1976 The geography of economic systems. Englewood Cliffs, N. J.: Prentice-Hall.

Berry, Brian J. L. and F. E. Horton
1970 Geographic perspectives on urban systems. Englewood Cliffs, N. J.: Prentice-Hall.

Black, William R.
1972 Interregional commodity flows: some experiments with the gravity model. Journal of Regional Science, 12,1: 107-118.

Black, William R.
1973 Toward a factorial ecology of flows. Economic Geography, 49:59-67.

Blalock, Hubert M., Jr.
1972 Social statistics. New York: McGraw-Hill.

Bohannan, Paul and Philip Curtin
1971 Africa and Africans. Garden City, N.Y.: Natural History Press. Quoted in Anthony de Souza and Philip W. Porter. The underdevelopment and modernization of the Third World:7. Washington, D.C.: Association of American Geographers, 1974.

Brinkmann, Theodor
1935 Theodor Brinkmann's economics of the farm business. Translated by E. T. Benedict, H. Stippler, and M. R. Benedict. Berkeley: University of California Press.

Britton, John N. H.
1965 The external relations of seaports. Tydschrift Voor Econ En Soc. Geografie, May-June:109-112.

1971 Methodology in flow analysis. The East Lakes Geographer, 7:22-36.

Bromley, R. J.
1971 Markets in the developing countries: a review. Geography, 56, part 2, no. 251:124-132.

Buckman, Harry O. and Nyle C. Brady
1968 The nature and properties of soils: a college text of edaphology. New York: The Macmillan Company.

Burdon, D. J.
1953 The underground water resources of Cyprus. Nicosia.

Cattell, Raymond B.
1965 Factor analysis: an introduction to essentials. I The purpose of underlying models. II The role of factor analysis in research. Biometrics, 21:190-215, 405-435.

1966 The Scree test for the number of factors. Multivariate Behavioral Research, 1:245-76.

Chisholm, Michael
1962 Rural settlement and land use. New York: John Wiley & Sons.

Christodoulou, Demetris
1959 The evolution of the rural land use pattern in Cyprus. Bude, Cornwall, England: Geographical Publications Limited.

Conkling, Edgar C. and Maurice Yeates
1976 Man's economic environment. New York: McGraw-Hill.

Cyprus
1946a A ten-year programme of development for Cyprus. Nicosia Nicosia: Government Printing Office.

1946b Census of population and agriculture. Nicosia: Government Printing Office.

Cyprus
1961 The rainfall of Cyprus. Technical Note 2. Nicosia: Ministry of Agriculture and Natural Resources, Meteorological Office.

1964 Cyprus population distribution 1960. Nicosia: Department of Lands and Surveys.

1965 Cyprus motor map. Nicosia: Department of lands and surveys.

1969a Cyprus topographical map. Nicosia: Department of Lands and Surveys.

1969b Statistical abstract 1968. Ministry of Finance, Department of Statistics and Research. Nicosia: Government Printing Office.

1970 Demographic report 1969. Ministry of Finance, Department of Statistics and Research. Nicosia: Government Printing Office.

1971a Cyprus town map series. Nicosia: Department of Lands and Surveys.

1971b Economic report 1970. Ministry of Finance, Department of Statistics and Research. Nicosia: Government Printing Office.

1971c Monthly statistics of quantity, price and value of vegetables and fruits brought to the municipal wholesale markets of Cyprus 1970. Nicosia: Ministry of Finance, Department of Statistics and Research.

1972a Agricultural survey 1971. Ministry of Finance, Department of Statistics and Research. Nicosia: Government Printing Office.

1972b Average annual precipitation map (1941-70) of Cyprus. Ministry of Agriculture and Natural Resources, Meteorological Service. Nicosia: Department of Lands and Surveys.

1972c Economic report 1971. Ministry of Finance, Department of Statistics and Research. Nicosia: Government Printing Office.

1972d Household survey 1971. Ministry of Finance, Department of Statistics and Research. Nicosia: Government Printing Office.

1973a Cadastral plans--sunprints. Nicosia: Department of Lands and Surveys.

1973b Cyprus administration and road map. Nicosia: Department of Lands and Surveys.

The Cyprus Gazette
1968 Wholesale market regulations. Supplement 3. Nicosia: Government Printing Office.

Cyprus Today
1967 Harnessing the rains of Cyprus. 5:10. Nicosia.

Darrah, L. B. and F. A. Tiongson
1969 Agricultural marketing in the Philippines. Laguna, Philippines: University of the Philippines College of Agriculture.

De Souza, Anthony and Philip W. Porter
1974 The underdevelopment and modernization of the Third World. Commission of College Geography, Resource Paper No. 28. Washington, D.C.: Association of American Geographers.

Dumont, René
1968 False start in Africa. London: Sphere Books Ltd.

1973 Socialisms and development. London: Deutsch.

Eyre, John D.
1959 Sources of Tokyo's fresh food supply. The Geographical Review, 49:455-474.

Fisher, W. B.
1971 The Middle East: a physical, social and regional geography. London: Methuen & Co.

Forman, Shepard and Joyce F. Riegelhaupt
1970 Market place and marketing system: toward a theory of peasant economic integration. Comparative Studies in Society and History, 12:188-212.

Goddard, J. B.
1970 Functional regions within the city center: a study by factor analysis of taxi flows in central London. Transactions, Institute of British Geographers, 49:161-182.

Good, Charles M.
1970 Rural markets and trade in East Africa: a study of the functions and developments of exchange institutions in Ankole, Uganda. Chicago: The University of Chicago, Department of Geography, Research Paper No. 128.

Greer-Wootten, Bryn
1972 A bibliography of statistical applications in geography. Geographic Technical Paper Series, Commission on College Geography, Technical Paper 9. Washington, D.C.: Association of American Geographers.

Gregor, Howard F.
1957 The local supply agriculture of California. Annals, Association of American Geographers, 47:267-276.

1970 Geography of agriculture: themes in research. Englewood Cliffs, New Jersey: Prentice Hall.

Hald, Marjorie W.
1968 A study of the Cyprus economy. Nicosia: Government Printing Office.

Hall, Peter, Ed.
1966 Von Thünen's isolated state. An English edition of Der Isolierte Staat by Johann Heinrich Von Thünen. Translated by Carla M. Wartenberg. Oxford: Pergamon Press.

Harman, Harry H.
1967 Modern factor analysis. Chicago: The University of Chicago Press.

Harris, Chauncy D. and Gerald A. P. Carrothers
1956 Historical review of the gravity and potential concepts in human interactance. Journal of the American Institute of Planners, 22:94-102.

Hay, Alan M.
1973 Transport for the space economy. London: Macmillan.

Helvig, Magne
1964 Chicago's external truck movements. Department of Geography Research Paper 90. Chicago: University of Chicago Press.

Hjiparaskevas, Leondios
1969 Report on the structure of marketing in the low coastal region of Paphos, the low coastal region of Limassol, the south-western region of Larnaca, and the Polis region. Ministry of Agriculture and Natural Resources, Cyprus Agricultural Research Institute. Nicosia: Government Printing Office.

Hodder, B. W. and Ukwu, U. I.
1969 Markets in West Africa: studies of markets and trade among the Yoruba and Ibo. Ibadan, Nigeria: Ibadan University Press.

Hunt, D. T.
1959 Market gardening in metropolitan Auckland. New Zealand Geographer, 15:130-155.

Isard, Walter
1960 Methods of regional analysis: an introduction to regional science. Cambridge, Mass.: The M.I.T. Press.

Jones, William O.
1972 Marketing staple food crops in tropical Africa. Ithaca, New York: Cornell University Press.

1974 Regional analysis and agricultural marketing research in tropical Africa: concepts and experience. Food Research Institute Studies, 13:3-28.

Jumper, Sidney R.
1964 The fresh vegetable industry in the U.S.A.: an example of dynamic inter-regional dependency. Tydschrift Voor Econ. En Soc. Geografie, 60:308-318.

Karouzis, G.
1970 Report on aspects of land tenure in Cyprus. Nicosia: Ministry of Agriculture and Natural Resources, Department of Agriculture.

Keefe, Eugene K. et. al.
1971 Area handbook for Cypurs. Washington, D.C.: U.S. Government Printing Office.

Kerlinger, Fred N. and Elazar J. Pedhazur
1973 Multiple regression in behavioral research. New York: Holt, Rinehart and Winston.

King, Leslie J.
1969 Statistical analysis in geography. Englewood Cliffs, New Jersey: Prentice-Hall.

Kohls, Richard and W. David Downey
1972 Marketing of agricultural products. Riverside, N.J.: Macmillan.

Mayer, Jean
1976 The dimensions of human hunger. Scientific American, 235,3:40-49.

Melamid, Alexander
1958 Land use in Cyprus. The Geographical Review, 47:113.

Mellor, John W.
1968 The economics of agricultural development. Ithaca, N.Y.: Cornell University Press.

Metcalf, David
1970 The economics of agriculture. Middlesex, England: Penguin.

Meyer, Albert Julius
1962 The economy of Cyprus. Cambridge: Harvard University Press.

Michaelides, R. C.
1970 Recent trends of agricultural production and productivity in Cyprus. Nicosia: Ministry of Agriculture and Natural Resources.

Mikesell, Marvin W.
1960 Market centers of northeastern Spain: a review. The Geographical Review, 50:247-251.

Morgan, W. B. and R. J. C. Munton
1971 Agricultural geography. London: Methuen.

Mosher, A. T.
1969 The development problems of subsistence farmers: a preliminary review. In Subsistence Agriculture and Economic Development, 6-11. Edited by Clifton R. Wharton, Jr. Chicago: Aldine.

Moyer, Reed and Stanley C. Hollander, eds.
1968 Markets and marketing in developing economies. First printing, prepared under the Sponsorship of the American Marketing Association (Chicago). Homewood, Illinois: Richard D. Irwin, Inc.

Mueller, John H. and Schuessler, Karl F.
1961 Statistical reasoning in sociology. Boston: Houghton Mifflin.

McMillan, Wendell M
1969 Development of agricultural marketing in the Middle East. In Stickley, Thomas S. (et. al., eds.) Man, Food and Agriculture in the Middle East, Symposium Proceedings, Faculty of Agricultural Sciences. Beirut: The American University of Beirut:65-80.

Nie, Norman H. et al.
1975 SPSS: statistical package for the social sciences. Second Edition. New York: McGraw Hill.

Norton, Ann and Richard Symanski
1975 The internal marketing systems of Jamaica. The Geographical Review, 65,4:461-475.

Olson, J. M.
1968 The spatial utilization of transportation modes in fruit and vegetable movements, 1961. Unpublished M.S. Thesis in Geography, University of Wisconsin.

Organization for Economic Cooperation and Development
1970 Food marketing and economic growth. Paris: OECD.

Panagides, Stahis Solomon
1967 An econometric study of the Cyprus economy. Ph.D. Dissertation, Department of Economics, Iowa State University of Science and Technology.

Papchristodoulou, Stelios
1970 Norm input-output data of the main crops of Cyprus. Nicosia: Cypurs Agricultural Research Institute.

Papasolomontos, A.
1972 Annual report of the Department of Agriculture. Ministry of Agriculture and Natural Resources, Department of Agriculture. Nicosia: Government Printing Office.

Patsalides, Andreas
1973 Address before the House of Representatives on the 1973 budget and the third five-year plan. Ministry of Finance. Nicosia: Government Printing Office.

Pierce, Jo Ann Bell
1975 Agricultural growth in an urban age. Gainsville, Florida: Institute of Food and Agricultural Sciences, University of Florida.

Rao, Potluri and Roger LeRoy Miller
1971 Applied econometrics. Belmont, California: Wadsworth.

Reed, Wallace E.
1967 Areal interaction in India, commodity flows of the Bengal-Bihar industrial area. Chicago: Department of Geography, University of Chicago, Research Paper No. 110.

Ricardo, David
1817 Principles of political economy and taxation. London: J. M. Dent & Sons.

Roman, James R.
1959 Transportation on Cyprus. Middle Eastern Affairs, 10: 367-371.

Rummel, R. J.
1967 Understanding factor analysis. Conflict Resolution, 11: 444-480.

1970 Applied factor analysis. Evanston: Northwestern University Press.

Schultz, Theodore W.
1964 Transforming traditional agriculture. New Haven, Conn.: Yale University Press.

Schwind, Paul J.
1975 A general field theory of migration: United States, 1955-1960. Economic Geography, 51,1:1-16.

Shepherd, Geoffrey S. and Gene A. Futrell
1970 Marketing farm products--economic analysis. Ames, Iowa: The Iowa State University Press.

Simmons, J. W.
1970 Interprovincial interaction patterns. The Canadian Geographer, 14:372-376.

Sinclair, Robert
1967 Von Thünen and urban sprawl. Annals, Association of American Geographers, 57:72-87.

Skinner, George W.
1964-5 Marketing and social structure in rural China. Journal of Asian Studies, XXIV:1-43, 195-228, 363-99.

Smith, Carol A.
1974 Economics of marketing systems: models from economic geography. Annual Review of Anthropology, 3:167-201.

Smith, D. L.
1966 Market gardening at Adelaide's urban fringe. Economic Geography, 42:19-36.

Smith, R. H. T.
1964 Toward a measure of complementarity. Economic Geography, 40:1-8.

1970 Concepts and methods in commodity flow analysis. Economic Geography, 46,2:404-416.

1972a Periodic markets in Africa, Asia, and Latin America. Monticello, Illinois: Council of Planning Librarians, Exchange Bibliography:318.

1972b Spatial structure and process in tropical West Africa. Economic Geography, 48,3.

Smith, Robert H. T. and Alan M. Hay
1969 A theory of the spatial structure of internal trade in underdeveloped countries. Geographical Analysis, 1,2: 121-136.

Smith, Vernon
1975 Marketing agricultural commodities in Pichincha Province, Ecuador. Geographical Review, 65,3:353-363.

Soteriades, C. G.
1970 Brief notes to accompany the reconnaissance soil map of Cyprus. Nicosia. (Mimeographed).

Soteriades, C. G. and G. C. Grivas
1961 Reconnaissance soil map of Cyprus. Ministry of Agriculture and Natural Resources, Soil Survey Section. Nicosia: Department of Lands and Surveys.

Soteriades, C. G. and G. C. Grivas
1970 General soil map of Cyprus. Ministry of Agriculture and Natural Resources. Nicosia: Department of Lands and Surveys.

South, Robert Browning
1972 An analysis of commodity flows and the spatial structure of the Bolivian economy. Unpublished Ph.D. dissertation, Department of Geography, University of Maryland.

Stöhr, Walter B.
1974 Interurban systems and regional economic development. Commission on College Geography, Resource Paper No. 26. Washington, D.C.: Association of American Geographers.

Symanski, Richard and R. J. Bromley
1974 Market development and the ecological complex. Professional Geographer, 26:382-388.

Thorp, Willard Long
1961 Cyprus: suggestions for a development programme. Prepared for the Government of the Republic of Cyprus. New York: United Nations.

Thünen, J. H. v.
1826 Der Isolierte Staat in Beziehung auf Landwirtschaft und Nationalökonomie. Hamburg: Perthes.

Toufexés, N. Chr. and J. Jacovides
1972 Hydrological year-book of Cyprus 1969-70. Nicosia: Ministry of Agriculture and Natural Resources, Department of Water Development.

Trager, Lillian
1976 Yoruba markets and trade: analysis of spatial structure and social organization in the Ijesaland marketing system. Unpublished Ph.D. thesis, University of Washington.

Tullstrom, N. H. O.
1970 Hydrogeological map. Geological Survey Department. Nicosia: Department of Lands and Surveys.

Ullmann, Edward L.
1957 American commodity flow: a geographical interpretation of rail and water traffic based on principles of spatial interchange. Seattle: University of Washington Press.

U. S. Department of Agriculture
1954 The yearbook of agriculture 1954, marketing. House Document No. 280, 83D Congress, Second Session. Washington, D.C.: Government Printing Office.

Van Nieuwenhuijze, C. A., ed.
1963 Markets and marketing as factors of development in the Mediterranean basin. Mediterranean Social Sciences Research Council, Second Assembly, Cairo, 1962. New York: Humanities.

Vance, James E., Jr.
1970 The merchants' world: the geography of wholesaling. Englewood Cliffs, N.J.: Prentice-Hall.

Went, Frits W.
1957 The experimental control of plant growth. Waltham, Mass.: The Chronica Botanica Company.

Whetham, Edith H.
1972 Agricultural marketing in Africa. London: Oxford University Press.

Whittlesey, D. W.
1936 Major agricultural regions of the world. Annals, Association of American Geographers, 26:199-240.

Wiseman, Robert
1976 Interpretation of factored flow data: a comparison of two traffic systems. The Professional Geographer, 28,2: 172-77.

Wortman, Sterling
1976 Food and agriculture. Scientific American, 235,3:31-39.

Yeates, Maurice
1974 An introduction to quantitative analysis in human geography. New York: McGraw-Hill.

Yiassemides, P. and H. Kunert
1967 Aspects of the agricultural economy of Cyprus 1950-1965. Misc. Publ. Cyprus. Agric. Res. Insti., No. 3. Nicosia: Ministry of Agricultural and Natural Resources.

Zipf, G. K.
1941 National unity and disunity. Bloomington, Indiana: Principia Press.

THE UNIVERSITY OF CHICAGO

DEPARTMENT OF GEOGRAPHY

RESEARCH PAPERS (Lithographed, 6×9 Inches)

(*Available from Department of Geography, The University of Chicago, 5828 S. University Ave., Chicago, Illinois 60637. Price: $6.00 each; by series subscription, $5.00 each.*)

106. SAARINEN, THOMAS F. *Perception of the Drought Hazard on the Great Plains* 1966. 183 pp.
107. SOLZMAN, DAVID M. *Waterway Industrial Sites: A Chicago Case Study* 1967. 138 pp.
108. KASPERSON, ROGER E. *The Dodecanese: Diversity and Unity in Island Politics* 1967. 184 pp.
109. LOWENTHAL, DAVID, et al. *Environmental Perception and Behavior.* 1967. 88 pp.
110. REED, WALLACE E. *Areal Interaction in India: Commodity Flows of the Bengal-Bihar Industrial Area* 1967. 210 pp.
112. BOURNE, LARRY S. *Private Redevelopment of the Central City: Spatial Processes of Structural Change in the City of Toronto* 1967. 199 pp.
113. BRUSH, JOHN E., and GAUTHIER, HOWARD L., JR. *Service Centers and Consumer Trips: Studies on the Philadelphia Metropolitan Fringe* 1968. 182 pp.
114. CLARKSON, JAMES D. *The Cultural Ecology of a Chinese Village: Cameron Highlands, Malaysia* 1968. 174 pp.
115. BURTON, IAN; KATES, ROBERT W.; and SNEAD, RODMAN E. *The Human Ecology of Coastal Flood Hazard in Megalopolis* 1968. 196 pp.
117. WONG, SHUE TUCK. *Perception of Choice and Factors Affecting Industrial Water Supply Decisions in Northeastern Illinois* 1968. 96 pp.
118. JOHNSON, DOUGLAS L. *The Nature of Nomadism* 1969. 200 pp.
119. DIENES, LESLIE. *Locational Factors and Locational Developments in the Soviet Chemical Industry* 1969. 285 pp.
120. MIHELIC, DUSAN. *The Political Element in the Port Geography of Trieste* 1969. 104 pp.
121. BAUMANN, DUANE. *The Recreational Use of Domestic Water Supply Reservoirs: Perception and Choice* 1969. 125 pp.
122. LIND, AULIS O. *Coastal Landforms of Cat Island, Bahamas: A Study of Holocene Accretionary Topography and Sea-Level Change* 1969. 156 pp.
123. WHITNEY, JOSEPH. *China: Area, Administration and Nation Building* 1970. 198 pp.
124. EARICKSON, ROBERT *The Spatial Behavior of Hospital Patients: A Behavioral Approach to Spatial Interaction in Metropolitan Chicago* 1970. 198 pp.
125. DAY, JOHN C. *Managing the Lower Rio Grande: An Experience in International River Development* 1970. 277 pp.
126. MAC IVER, IAN. *Urban Water Supply Alternatives: Perception and Choice in the Grand Basin, Ontario* 1970. 178 pp.
127. GOHEEN, PETER G. *Victorian Toronto, 1850 to 1900: Pattern and Process of Growth* 1970. 278 pp.
128. GOOD, CHARLES M. *Rural Markets and Trade in East Africa* 1970. 252 pp.
129. MEYER, DAVID R. *Spatial Variation of Black Urban Households* 1970. 127 pp.
130. GLADFELTER, BRUCE. *Meseta and Campiña Landforms in Central Spain: A Geomorphology of the Alto Henares Basin* 1971. 204 pp.
131. NEILS, ELAINE M. *Reservation to City: Indian Urbanization and Federal Relocation* 1971. 200 pp.
132. MOLINE, NORMAN T. *Mobility and the Small Town, 1900–1930* 1971. 169 pp.
133. SCHWIND, PAUL J. *Migration and Regional Development in the United States, 1950–1960* 1971. 170 pp.
134. PYLE, GERALD F. *Heart Disease, Cancer and Stroke in Chicago: A Geographical Analysis with Facilities Plans for 1980* 1971. 292 pp.
135. JOHNSON, JAMES F. *Renovated Waste Water: An Alternative Source of Municipal Water Supply in the U.S.* 1971. 155 pp.
136. BUTZER, KARL W. *Recent History of an Ethiopian Delta: The Omo River and the Level of Lake Rudolf* 1971. 184 pp.
137. HARRIS, CHAUNCY D. *Annotated World List of Selected Current Geographical Serials in English, French, and German* 3rd edition 1971. 77 pp.
138. HARRIS, CHAUNCY D., and FELLMANN, JEROME D. *International List of Geographical Serials* 2nd edition 1971. 267 pp.
139. MC MANIS, DOUGLAS R. *European Impressions of the New England Coast, 1497–1620* 1972. 147 pp.
140. COHEN, YEHOSHUA S. *Diffusion of an Innovation in an Urban System: The Spread of Planned Regional Shopping Centers in the United States, 1949–1968* 1972. 136 pp.

141. MITCHELL, NORA. *The Indian Hill-Station: Kodaikanal* 1972. 199 pp.

142. PLATT, RUTHERFORD H. *The Open Space Decision Process: Spatial Allocation of Costs and Benefits* 1972. 189 pp.

143. GOLANT, STEPHEN M. *The Residential Location and Spatial Behavior of the Elderly: A Canadian Example* 1972. 226 pp.

144. PANNELL, CLIFTON W. *T'ai-chung, T'ai-wan: Structure and Function* 1973. 200 pp.

145. LANKFORD, PHILIP M. *Regional Incomes in the United States, 1929–1967: Level, Distribution, Stability, and Growth* 1972. 137 pp.

146. FREEMAN, DONALD B. *International Trade, Migration, and Capital Flows: A Quantitative Analysis of Spatial Economic Interaction* 1973. 202 pp.

147. MYERS, SARAH K. *Language Shift Among Migrants to Lima, Peru* 1973. 204 pp.

148. JOHNSON, DOUGLAS L. *Jabal al-Akhdar, Cyrenaica: An Historical Geography of Settlement and Livelihood* 1973. 240 pp.

149. YEUNG, YUE-MAN. *National Development Policy and Urban Transformation in Singapore: A Study of Public Housing and the Marketing System* 1973. 204 pp.

150. HALL, FRED L. *Location Criteria for High Schools: Student Transportation and Racial Integration* 1973. 156 pp.

151. ROSENBERG, TERRY J. *Residence, Employment, and Mobility of Puerto Ricans in New York City* 1974. 230 pp.

152. MIKESELL, MARVIN W., editor. *Geographers Abroad: Essays on the Problems and Prospects of Research in Foreign Areas* 1973. 296 pp.

153. OSBORN, JAMES. *Area, Development Policy, and the Middle City in Malaysia* 1974. 273 pp.

154. WACHT, WALTER F. *The Domestic Air Transportation Network of the United States* 1974. 98 pp.

155. BERRY, BRIAN J. L., et al. *Land Use, Urban Form and Environmental Quality* 1974. 464 pp.

156. MITCHELL, JAMES K. *Community Response to Coastal Erosion: Individual and Collective Adjustments to Hazard on the Atlantic Shore* 1974. 209 pp.

157. COOK, GILLIAN P. *Spatial Dynamics of Business Growth in the Witwatersrand* 1975. 143 pp.

158. STARR, JOHN T., JR. *The Evolution of Unit Train Operations in the United States: 1960–1969—A Decade of Experience* 1976. 247 pp.

159. PYLE, GERALD F. *The Spatial Dynamics of Crime* 1974. 220 pp.

160. MEYER, JUDITH W. *Diffusion of an American Montessori Education* 1975. 109 pp.

161. SCHMID, JAMES A. *Urban Vegetation: A Review and Chicago Case Study* 1975. 280 pp.

162. LAMB, RICHARD. *Metropolitan Impacts on Rural America* 1975. 210 pp.

163. FEDOR, THOMAS. *Patterns of Urban Growth in the Russian Empire during the Nineteenth Century* 1975. 275 pp.

164. HARRIS, CHAUNCY D. *Guide to Geographical Bibliographies and Reference Works in Russian or on the Soviet Union* 1975. 496 pp.

165. JONES, DONALD W. *Migration and Urban Unemployment in Dualistic Economic Development* 1975 186 pp.

166. BEDNARZ, ROBERT S. *The Effect of Air Pollution on Property Value in Chicago* 1975. 118 pp.

167. HANNEMANN, MANFRED. *The Diffusion of the Reformation in Southwestern Germany, 1518-1534* 1975. 248 pp.

168. SUBLETT, MICHAEL D. *Farmers on the Road. Interfarm Migration and the Farming of Noncontiguous Lands in Three Midwestern Townships, 1939-1969* 1975. 228 pp.

169. STETZER, DONALD FOSTER. *Special Districts in Cook County: Toward a Geography of Local Government* 1975. 189 pp.

170. EARLE, CARVILLE V. *The Evolution of a Tidewater Settlement System: All Hallow's Parish, Maryland, 1650–1783* 1975. 249 pp.

171. SPODEK, HOWARD. *Urban-Rural Integration in Regional Development: A Case Study of Saurashtra, India—1800–1960* 1976. 156 pp.

172. COHEN, YEHOSHUA S. and BERRY, BRIAN J. L. *Spatial Components of Manufacturing Change* 1975 272 pp.

173. HAYES, CHARLES R. *The Dispersed City: The Case of Piedmont, North Carolina* 1976. 169 pp.

174. CARGO, DOUGLAS B. *Solid Wastes: Factors Influencing Generation Rates* 1977. 112 pp.

175. GILLARD, QUENTIN. *Incomes and Accessibility. Metropolitan Labor Force Participation, Commuting, and Income Differentials in the United States, 1960–1970* 1977. 140 pp.

176. MORGAN, DAVID J. *Patterns of Population Distribution: A Residential Preference Model and Its Dynamic* 1978 216 pp.

177. STOKES, HOUSTON H.; JONES, DONALD W. and NEUBURGER, HUGH M. *Unemployment and Adjustment in the Labor Market: A Comparison between the Regional and National Responses* 1975. 135 pp.

178. PICCAGLI, GIORGIO ANTONIO. *Racial Transition in Chicago Public Schools. An Examination of the Tipping Point Hypothesis, 1963–1971* WITHDRAWN.

179. HARRIS, CHAUNCY D. *Bibliography of Geography. Part I. Introduction to General Aids* 1976. 288 pp.

180. CARR, CLAUDIA J. *Pastoralism in Crisis. The Dasanetch and their Ethiopian Lands.* 1977. 339 pp.

181. GOODWIN, GARY C. *Cherokees in Transition: A Study of Changing Culture and Environment Prior to 1775.* 1977. 221 pp.

182. KNIGHT, DAVID B. *A Capital for Canada: Conflict and Compromise in the Nineteenth Century.* 1977. 359 pp.

183. HAIGH, MARTIN J. *The Evolution of Slopes on Artificial Landforms: Blaenavon, Gwent.* 1978. 311 pp.

184. FINK, L. DEE. *Listening to the Learner. An Exploratory Study of Personal Meaning in College Geography Courses.* 1977. 200 pp.

185. HELGREN, DAVID M. *Rivers of Diamonds: An Alluvial History of the Lower Vaal Basin.* 1979.

186. BUTZER, KARL W., *editor. Dimensions of Human Geography: Essays on Some Familiar and Neglected Themes.* 1978. 201 pp.

187. MITSUHASHI, SETSUKO. *Japanese Commodity Flows.* 1978. 185 pp.

188. CARIS, SUSAN L. *Community Attitudes toward Pollution.* 1978. 226 pp.

189. REES, PHILIP M. *Residential Patterns in American Cities, 1960.* 1979. 424 pp.

190. KANNE, EDWARD A. *Fresh Food for Nicosia.* 1979. 116 pp.